THE AIRBORNE MAFIA

A volume in the series

Battlegrounds: Cornell Studies in Military History

Edited by David J. Silbey

Editorial Board: Adela Cedillo, M. Girard Dorsey, Michael W. Hankins, Ellen D. Tillman, and Edward B. Westermann

A list of titles in this series is available at cornellpress.cornell.edu.

THE AIRBORNE MAFIA

THE PARATROOPERS WHO SHAPED AMERICA'S COLD WAR ARMY

ROBERT F. WILLIAMS

CORNELL UNIVERSITY PRESS
Ithaca and London

First published 2025 by Cornell University Press

Printed in the United States of America

Library of Congress Cataloging-in-Publication Data

Names: Williams, Robert F. (Ph. D. in history), author.
Title: The airborne mafia : the paratroopers who shaped America's Cold War army / Robert F. Williams.
Description: Ithaca : Cornell University Press, 2025. | Series: Battlegrounds: Cornell studies in military history | Includes bibliographical references and index.
Identifiers: LCCN 2024028895 (print) | LCCN 2024028896 (ebook) | ISBN 9781501779824 (hardcover) | ISBN 9781501779831 (ebook) | ISBN 9781501779848 (pdf)
Subjects: LCSH: United States. Army—Parachute troops—History—20th century. | United States. Army—Officers—History—20th century. | Organizational behavior—United States. | Airborne operations (Military science)—History—20th century.
Classification: LCC UG633 .W5676 2025 (print) | LCC UG633 (ebook) | DDC 356/.166097309045—dc23/eng/20240809
LC record available at https://lccn.loc.gov/2024028895
LC ebook record available at https://lccn.loc.gov/2024028896

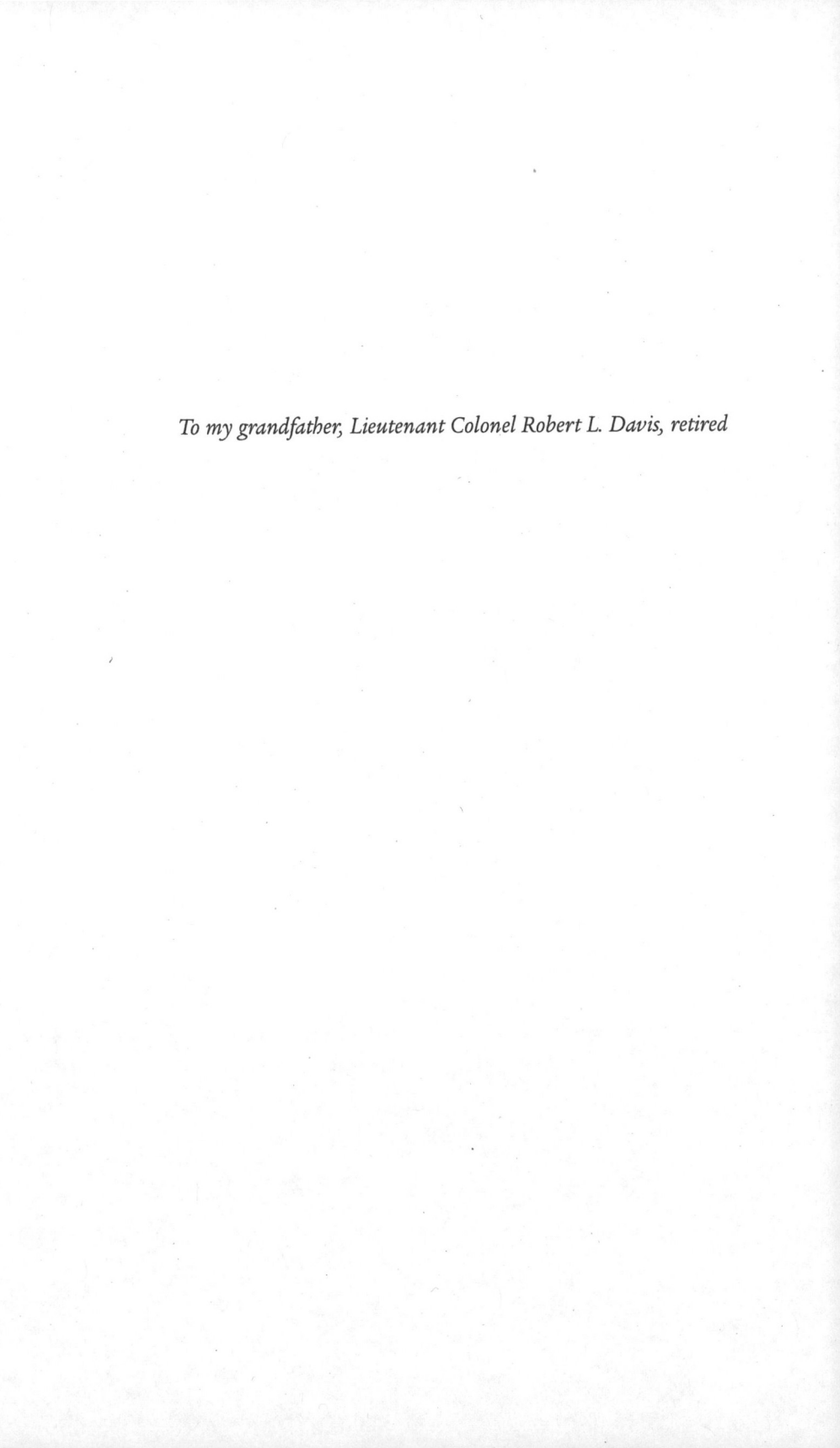

To my grandfather, Lieutenant Colonel Robert L. Davis, retired

CONTENTS

ILLUSTRATIONS

Preface

Little groups of paratroopers (LGOPs) seem to run the army. They certainly do in the airborne. This phenomenon stems from a cultural expectation that because decentralized autonomous groups of varying ranks accomplished their missions during World War II, they will continue to do so in the twenty-first century. This attitude affects all activities in the 82nd Airborne Division, from planning for a high-intensity live-fire range to cleaning disparate areas on-base for the annual Operation Clean Sweep. In my experience, commanders issue guidance, but execution is decentralized in all facets of daily operations. This gives paratroopers a swagger extending well beyond the drop zone and into the everyday operations of airborne units. It is also carried throughout the army.

This book is drawn from my decade on jump status across multiple airborne units. I was immersed in the culture and history of these elite forces. The idea for this project germinated, really, twice. First in 2006 when I arrived in Italy at my first airborne assignment and was gifted a copy of E. M. Flanagan's *Airborne* so that I could better understand the history of airborne units. And again, in 2011 when while deployed to Khost Province, Afghanistan, I found a copy of James M. Gavin's memoir, *On to Berlin*. I read the book voraciously during that tour of duty. It was my second airborne unit, and I was already steeped in the culture, but reading this introduced me to one of the most charismatic leaders in army history. I did not know it then, but that started me on this journey into the history of airborne culture, with Gavin naturally at its center. That copy is still on my shelf, more marked up than any of the memoirs I own.

Later, I spent the seventieth D-Day anniversary in Sainte-Mère-Église in Normandy, witnessing the legacy of that day while jumping onto a drop zone next to La Fière bridge. I went to jumpmaster school and obtained master parachutist and Pathfinder badges. I loved jumping from planes, and that love motivated my research into the attitudes still shaping airborne units today. I uncover these cultural roots in the creation of the first airborne divisions. Their innovations in tactics and technology shaped an ethos passed through generations of paratroopers. This outlook influenced operations, fighting spirit, and self-perception as an elite force. By documenting airborne history from the start, I contextualize contemporary attitudes, situating today's paratrooper within an evolving legacy that informs everything from rituals to cohesion. My book illuminates a culture that shaped my own path in and out of uniform.

This book is personal but does not center on my experience. Its origins are in my experience as both an observer and a product of airborne culture, but it is about the origins of that culture in which I took part. Initially I wanted to study culture at the soldier level, which comes out at times. But as historians realize, sources drive you in different ways. As research and writing progressed, I realized this was less about how young soldiers viewed airborne service and more about how pioneering a new battlefield arrival method had lasting effects on the US Army. Though rooted in my own history, this work is the product of careful study of the World War II and early Cold War US Army. I've dedicated this book to all paratroopers who have lived this airborne culture.

Acknowledgments

Nobody writes a book alone. From mentors to inspiration, to help with the minutia of research or life, this has been a team effort through and through. I am simply the lucky one able to put my name on the cover. I must first acknowledge all those whom I have had the honor of serving with throughout my career in the US Army, and especially those we lost in Iraq and Afghanistan; you are dearly missed. I hope that the effort that went into this book serves as an example of how to harness our experiences into new purpose and meaning. Many thanks also to Kyle Jahner and Marc DeVore, who provided the initial impetus for this study. That was the first time the *Army Times* has motivated me to do more than throw the magazine away.

My doctoral adviser, Peter Mansoor, has been a mentor par excellence. He provided an environment much like that I was used to in the Army, a classic officer–noncommissioned officer mentorship relationship, for which I am eternally grateful. Without his assistance and support, along with that of Geoffrey Parker, Lydia Walker, Mark Grimsley, David Steigerwald, Jennifer Siegel, R. Joseph Parrott, Randolph Roth, Christopher McKnight Nichols, Bartow Elmore, and the rest of the History Department faculty, this book would not have happened. Likewise, my Buckeye graduate school peers have been critical to my development as a historian: Ben St. Angelo, Max von Bargen, Rebecka Beard, Mitch Carter, Hugh "Buck" Gardenier, Benjamin Lyman, Paul McAllister, Carson Teuscher, and Guido Rossi. We made graduate school a "team sport," and if it were not for a global pandemic, it might have been even more fun.

Also, thanks go to many outside Dulles Hall. My intellectual journey toward this project began at the University of North Carolina at Chapel Hill, and I owe a debt of gratitude to the many encouraging scholars I met there. I thank my professors in Chapel Hill, Joseph Glatthaar, Wayne Lee, and Miguel La Serna, for inspiring me to pursue graduate-level history. Mary Elizabeth Walters, Joseph Stieb, and J. Davis Winkie were incredible graduate student mentors there as well. Of course, the folks who helped ease my transition from paratrooper to academic go far beyond the historians I mention. Dana Copeland helped start me off on the right foot in the summer of 2017. Hilary Lithgow, professor of English at Chapel Hill, has provided invaluable guidance, mentorship, and coaching to develop my writing skills. Likewise, she is an invaluable resource for veterans making the transition to university life at Carolina. Without her, Shane Hale, and Luke Fayard, Chapel Hill might have been a dreary place for veterans. But most significantly, I must thank my dear friend Eric Michael Burke, who showed me that, yes, an enlisted grunt can do serious history.

Many others outside the campuses in Columbus and Chapel Hill have been incredible in so many ways. Ingo Trauschweizer and I had essential early conversations that shaped the research for this manuscript. Cavender Sutton, Adam Toering, Nate Finney, John Curatola, Ben Schneider, and Ruth Lawlor have all read chapters or assisted with the source material to make this work even better. Led by Gregory Daddis, Lorien Foote, Rob Citino, and Kara Dixon Vuic, the 2022 summer seminar in military history provided an inspiring group within the broader military history community that I lean on consistently. Steph Hinnershitz, Adam Givens, Jennifer Popowycz, Mike Bell, and Jeremy Collins, alongside fellow "Camp Clausewitz" campers Ryan Booth, Marjorie Galelli, Seth Givens, Hayley Hasik, Chris Juergens, Shaun Mawdsley, James Sandy, Ryan Reynolds, and the rest, have been vital sources of mentorship, advice, and sometimes emotional support.

The research for this book has been supported by generous funding from all over the country. Thanks to the Department of History at Ohio State University, the Mershon Center for International Security Studies, the Bradley Foundation, the Society for Military History, the Eisenhower Foundation, and the Truman Foundation for providing me with generous travel grants. Without this tremendous support, this book would not have been possible. The Mershon Center especially provided me with a second academic home at Ohio State. Dorry Noyes runs a tight ship, and her team of Kyle McCray, Andrew Mackey, and Dani Wollerman helped me grow as an academic and person while on campus.

Many thanks also to my colleagues at Fort Leavenworth at Army University Press. Especially to Roy Parker, who despite his cavalry nature, welcomed this old paratrooper with open arms, and alongside Donald Wright, Kate Dahlstrand, Randy Masten, Kevin Kennedy, Rob

Thompson, Chris Carey, and so many others helped me find not only an intellectual home but a cultural one as well. I could not ask for a better job after making the probably ill-advised decision to go to graduate school.

No historical project is possible without those admirable archivists who "know where the bodies are buried." The wonderful team within the National Archives and Presidential Library system, as well as the Library of Congress, were critical for digging through the vast army records system. The Bentley Historical Library and Anderson University Library were instrumental in getting me started while the large collections were closed during the pandemic. Without the assistance of Diana Bachman, Sarah McLusky, and Caitlin Moriarty at Michigan and Nicholas Stanton-Roark at Anderson, this work would have taken at least another year. Many thinks must also go to Abigail Gardner at the National Defense University Library; John Aarsen, Rafael Alvarez, and Chris Ruff at the 82nd Airborne Museum; Susan Lintelmann at the United States Military Academy Library; and Justine Melone and Thomas Buffenbarger at the US Army Heritage and Education Center.

I am deeply indebted to David Silbey, the series editor, and Bethany Wasik, my acquisitions editor at Cornell. They were instrumental in seeing my vision and helping bring it to fruition. They exhibited unfailing confidence in my project from the first time we spoke, a confidence I did not have in myself. Also, many thanks to Maysan Haydar, my fellow Buckeye and Mansoor advisee who finally introduced me to David. The two unnamed peer reviewers provided indispensable feedback that I endeavored to incorporate to make this book even better. I am humbled and honored to receive such amazing help from them and from the rest of my colleagues who have read this manuscript in full or in part. And, of course, my gratitude goes to the entire team at Cornell University Press, including Kristen Gregg and Karen Laun for helping me through the publication process. And to Glenn Novak, for taking a decent manuscript and making it shine. Thank you for helping make my dream of publishing a book become a reality.

But most of all, I must thank my family. My amazing parents, Fred and Debbie Williams, provided an essential foundation upon which to build my mind and were models for both parenthood and work ethic. My father passed away in 2003, and though he is the epitome of the working class and did not attend college, I am sure he is more than proud. My grandfather, who served as an officer in the US Army from 1943 to 1964, the primary years that this study concerns itself with, was likewise an inspiration for me to understand better the period in which he dedicated his life to service. With everyone blazing the trail, both figuratively and literally, this accomplishment is possible. And finally, my wife and children: Megan, Charlotte, and Franklin. Long days and nights pursuing this project have often made life difficult at home. Still, your encouragement and love have been nothing short of instrumental in completing this project. I love you and the kids more than I could ever express.

Any errors and shortcomings contained within the pages that follow remain the sole responsibility of the author. My sincere hope is that this book makes everyone proud, especially after having played such a vital role in this process. But most important are those to whom the book is dedicated.

Abbreviations

ACoS	assistant chief of staff
AGF	Army Ground Forces
AIR	airborne infantry regiment
AR	Army regulation
ATFA–1	Atomic Field Army–1
CGSC	Command and General Staff College
CIA	Central Intelligence Agency
CINCFE	Commander in chief, Far East
CONARC	Continental Army Command
CoS	chief of staff
DA	Department of the Army
DoD	Department of Defense
DZ	drop zone
FM	field manual
FSR	field service regulations
G-1	personnel
G-2	intelligence
G-3	operations
G-4	logistics
GIR	glider infantry regiment
ICBM	intercontinental ballistic missile
IRBM	intermediate-range ballistic missile
JCS	Joint Chiefs of Staff
LZ	landing zone
MACV	Military Assistance Command–Vietnam
MOMAR	Modern Mobile Army
NATO	North Atlantic Treaty Organization
NCO	noncommissioned officer
NDA	National Defense Act
NLF	National Liberation Front

NSC	National Security Council
OSD	Office of the Secretary of Defense
PAVN	People's Army of Vietnam
PIR	parachute infantry regiment
ROAD	Reorganization Objective Army Division
ROCAD	Reorganization of the Current Armored Division
ROCID	Reorganization of the Current Infantry Division
ROK	Republic of Korea
ROTAD	Reorganization of the Airborne Division
SAC	Strategic Air Command
SAMs	surface-to-air missiles
SHAEF	Supreme Headquarters, Allied Expeditionary Forces
SSMs	surface-to-surface missiles
STRAC	Strategic Army Corps
STRAF	Strategic Army Forces
TO&E	table of organization and equipment
WSEG	Weapons Systems Evaluation Group

THE AIRBORNE MAFIA

Introduction
An Airborne Culture

> And where is the Prince who can afford so to
> cover his Country with Troops for its Defense,
> as that Ten Thousand Men descending from the
> Clouds, might not in many Places do an infinite
> deal of Mischief, before a Force could be brought
> together to repel them?
>
> —Benjamin Franklin, 1784

At 6:47 p.m. on September 19, 1994, the first of 113 American C-130 Hercules and C-141 Starlifter aircraft lifted off from the runway at Pope Air Force Base in North Carolina. These airplanes were loaded with combat-equipped members of the 82nd Airborne Division en route to Haiti for Operation Uphold Democracy. The drop was to be the largest parachute drop since World War II. Except it never happened. The planes left with paratroopers ready to make a combat jump that night but were recalled seventy-three minutes into their flight—halfway to their drop zones. Former president Jimmy Carter and former chairman of the Joint Chiefs of Staff Colin Powell's peace talks succeeded after Haitian general Raoul Cedras learned that the American paratroop invasion force was headed to the island. Carter later credited Haitian knowledge that the 82nd Airborne Division was on its way as critical to reaching a settlement. Cedras would not begin serious negotiations until he confirmed that the 82nd was in the air. The mere thought of thousands of parachutists descending from the clouds upon Haiti was enough for cooler heads to prevail, and the American ability to project overwhelming force helped prevent a bloody invasion.[1]

Twenty-seven years later, 82nd Airborne Division commander Maj. Gen. Christopher Donahue was the last American service member to step off Afghan soil. He was completing the latest quick-response mission asked of his division and embodied the expeditionary culture that evolved within airborne units in the United States since 1940. The division's Immediate Response Force, including its division headquarters, was sent to Kabul to help assume control of airfield security as the situation deteriorated at Hamid Karzai International Airport during the largest noncombatant evacuation operation in American history. Nevertheless, when the green-hued night-vision image of Donahue stepping onto the last C-17 Globemaster III aircraft on August 31, 2021, went viral, he was reinforcing the culture of his division and the airborne: a culture of readiness and decentralization that evolved over more than eighty years to inspire ingenuity and initiative, where privates and major generals share hardship, and "leaders jump first and eat last."[2]

This culture originated in the development of airborne units during the early stages of World War II, was reinforced by wartime experiences, and had an enormous impact on the Cold War US Army. Key leaders from airborne units controlled the army's direction in the 1950s. Their ties to one another and the press brought rapid promotions and prominence as the service grasped for relevance in the air-atomic age. They dominated the strategic and tactical thinking of the army and made recommendations and changes based on their experiences with airborne warfare in World War II. The airborne mafia, as they would come to be called, then ushered in organizational changes centered on decentralization and mobility, helicopter-borne airmobile tactics, and a strategic response force that provided the army with a rapidly deployable force projection capability from within the United States. These key leaders did not always agree, yet their ideas—often divergent from one another—were the product of their shared understanding of military operations filtered through their wartime experience leading airborne units. This is the story of how one group of leaders created a subculture that permeated the entire army. Their rise to prominence allowed these officers and their subordinates to enact institutional changes that reflected their ideas about how to fight. The airborne mafia imbued the army with an air-minded expeditionary mindset that has impacted how the army has organized itself and fought into the twenty-first century.

The "airborne mafia" refers to the cadre of World War II airborne officers who took control of the US Army in the postwar years. These officers have been referred to by several nicknames, including "the parachute club," "the airborne club," but, most memorably, the airborne mafia. While the term "mafia" connotes an organized crime syndicate, in this case it refers to a group of like-minded people with a shared background who helped and protected one another, sometimes to the disadvantage of other groups. This group controlled the institutional direction of the US Army from the aftermath of Korea into the Vietnam War. This book centers on three leading officers: the airborne triumvirate of Matthew B. Ridgway, Maxwell D. Taylor, and James M. Gavin. It details their efforts in crafting a new way of fighting, leading airborne units in World War II, and their subsequent postwar careers diffusing their learned cultural behavior around the army. Operational behaviors and ideas that originated in the Second World War about the efficacy of small-unit tactics fueled the development of tactical mobility and contingency forces tailor-made to fight low-intensity Cold War conflicts. While these officers did not always agree, their postwar thinking was deeply influenced by their shared experiences leading airborne units in combat during the war.[3] Soldiers from all army branches came together to form the airborne. This mixing of different specialties led the airborne units to develop a shared mindset and subculture. The airborne was unique in that all parachute units traced their origin back to the original test platoon formed in 1940. As the airborne expanded, the test platoon veterans became the core cadre of the first parachute battalion. Then, members of that battalion trained successive battalions, regiments, and divisions, thus passing on the airborne culture. This cadre system meant new airborne units emerged with the same norms, values, and attitudes as the test platoon where it all started.

Airborne units operated under several critical cultural tenets. First, an attitude of exceptionalism and unit pride facilitated self- and collective confidence vital for a group that experienced isolation as a normal battlefield condition. Second, flexibility, innovative thinking, and adaptability combined with a streamlined unit structure to create an expeditionary mindset and ability to respond to fluid battlefield scenarios at the tactical level. Innovative and adaptable minds were also required to solve complex problems, sometimes only discovered after committing airborne forces to combat. Third, an air-minded expeditionary approach fueled notions of vertical envelopment in all its conceptions—parachute, glider, assault transport, and helicopter. An expeditionary ethos, like that which emerged within the airborne, is an innovative and adaptable mindset that allows units to respond with minimal preparation time to a broad array of missions. It also involves streamlining unit structures and equipment for maximum air transportability.[4] Finally, decentralization and individuality were crucial for a unit that must fight dispersed and maintain the initiative, often without high-command supervision. Trust in junior leaders was essential to overcoming the dispersion inherent in parachute operations in its burgeoning stage. These ideas were unique insofar as airborne soldiers saw the rest of the army as rule-bound and conventional. Taken together, these ideas formed the shared experiences that cemented relationships and established the unit cohesion that paid dividends in combat and beyond.

These cultural tenets arose from basic underlying assumptions about the nature of airborne warfare. Early airborne divisions contained significantly less organic—permanently assigned—ground transportation capability than other army divisions. They were explicitly designed to be more lightly equipped than standard infantry units. As units grew to regiment and division size, the assumption remained that they would operate without support assets enjoyed by regular infantry divisions—especially armor and other heavy weapons. Paratroopers also realized that massive dispersal would be expected and that they would find themselves fighting in small, ad hoc units with a minimum of formal leaders present. In the postwar environment, this culture manifested itself in the army's confidence in operational reach and flexibility, battlefield decentralization, and overall air-mindedness. Paratroopers became a critical component of American strategic culture during the Cold War, as airborne leaders and the mystique surrounding their exploits during World War II helped change ideas and patterns of behavior within the national strategic community. But there was not a uniformity of thought; the airborne mafia consisted of strong personalities who often disagreed with one another. Yet the combination of their ideas, their prewar experiences, and the shared experience of airborne combat in World War II shaped how they saw the use of land warfare during the early Cold War.

As ostensibly elite units during World War II, the airborne collected the best personnel—volunteers who scored higher on entrance and physical examinations upon entering the army and who received extra pay for assuming the risk inherent in parachute operations. The airborne also benefited from higher-quality leaders, as dangerous parachute duty attracted some of the army's best officers and offered them quicker promotions thanks to their elite status. This also included officers who might have ascended to the highest general officer ranks regardless of airborne service. Still, the glamour and postwar prestige that came with jump wings on their uniform helped propel them forward at an accelerated rate. This talent uptake led to many former paratroopers ascending to prominence in the postwar army. It is conceivable that these officers would have reached similar career heights even without their service in airborne units; nevertheless, airborne units became famous for their exploits throughout the war, and the postwar army had proportionally more airborne divisions than it had during the war. This propelled the airborne mindset to prominence in the army during the Cold War, and some leaders considered applying airborne ideas to the rest of the army.[5]

Maj. Gen. William C. Lee pioneered early airborne units, earning the title "father of the airborne," but Ridgway emerged as the patriarch of the airborne mafia. Ridgway's two key subordinates, Maxwell Taylor and James Gavin, served under him in the 82nd Airborne Division early on. Gavin later commanded the 82nd, while Taylor led the 101st Airborne in the war. Though these three men had differing personalities and leadership styles, all had formative shared experiences in the 82nd, forging bonds as the first airborne generals. The airborne mafia consisted of many beyond these top leaders. It included a broader network of officers they mentored, who embraced their vision for the army's future based on airborne innovations and culture. One significant addition during the post-Korea years was Hamilton Howze, who brought the spirit of the old horse cavalry to fledgling air mobility projects in the mid-1950s.

That the airborne mafia came to run the army was far from preordained. Yet, as they ascended to the highest ranks of the army, they transmitted their culture in four significant ways. First, their comfort with ushering in progressive ideas about warfare and fighting behind enemy lines manifested itself in the changing notions of strategy and civil-military relations in the Eisenhower and Kennedy administrations. Second, an emphasis on decentralization and command of multiple units on a nebulous, noncontiguous battlefield influenced what would become the "pentomic" doctrine and unit design of the army for maneuvering on the hypothetical nuclear battlefield. The term "pentomic" referred to its five-sided design and atomic capability. Third, the development of the airmobile division borrowed heavily from airborne tactics and air-mindedness to provide the tactical battlefield mobility that airborne units lacked during World War II. Fourth, the advent of strategic response forces was the result of similar quick-response-force missions executed during the invasion of Italy and the reinforcement of Allied forces during the Ardennes counteroffensive of

December 1944 and a simultaneous emphasis on the aerial delivery of combat power as experienced during World War II. The book explores how this distinct culture developed before the war, was forged in combat, and then influenced the Cold War army as airborne generals rose in rank. It is the story of how an innovative subculture changed the broader institution.[6]

The outlook and priorities of the airborne subculture significantly shaped military strategy and policy. While the airborne leaders each had his own views, these were grounded in the shared values forged fighting together in World War II. Many factors drove institutional change in the 1950s and '60s—the Soviet threat, Allied forces, budget woes, and new technology. However, the airborne mafia imparted their values and beliefs widely across the army, giving national leaders new perspectives. They advocated ideas to fight so-called limited wars, like rapid-response intervention forces and units ready for limited nuclear combat. These ideas reflected their understanding of war on a spectrum. They bolstered the army's case for increased funding amid debates about whether airpower or ground troops were more useful in future combat. The airborne mafia's imprint on the army remains today. This subculture created lasting changes by stressing flexibility, decentralization, and delivering combat power by air. Its wartime experiences diffused widely, shaping the institution.

The United States is, of course, not the only great power to maintain airborne forces after World War II, and likewise not the only one to experience the dominating influence of elite paratroop officers. In the French army, the paratroop and Foreign Legion units that fought in Indochina and Algeria significantly impacted the French army and French civil-military relations. French paratroopers enjoyed near cult-like status, threatened a parachute drop on Paris, and are credited with orchestrating the coup that helped return Charles de Gaulle to office. French paratroopers pioneered the widespread use of helicopters and air mobility in Algeria. The Soviet airborne forces featured prominently in postwar planning, became a veritable separate service after World War II, and served as imperial storm troopers for the Soviet Union. Paratroopers enjoyed less institutional pull in the United Kingdom but remained prominent for force projection into former colonial holdings. While large-scale parachute assaults against determined enemy forces had reached their nadir in 1945, the romanticism of tough, well-trained, elite units prepared to jump behind enemy lines remained prominent in these four countries.[7]

Culture is the central organizing principle of this book. Culture is both the bedrock of military effectiveness and a critical concept for understanding change in military organizations. Organizational culture is shared assumptions, values, and beliefs that influence how people behave in the military, a service branch, a unit, and so on. Its impact on operations and, by proxy, effectiveness is inherent. It is a learned way of thinking about work and operating collectively to pursue group objectives. To analyze culture, one must examine its three components: artifacts, espoused values, and beliefs, as well as basic underlying assumptions.[8]

Artifacts are the most visible parts of a culture to evaluate—the visible and tangible phenomena. These include routines and rituals, celebrations, unique clothing, manners of address, doctrine, and the myths and stories an organization tells about itself. In the airborne, these included wearing jump wings and jump boots and using special cadences for marching and jogging. Values are the expressed, profound, normative convictions that determine which types of behavior the group desires. In World War II airborne units, values such as flexibility, initiative, and leadership were expressly sought within parachute units. Beliefs represent a group's convictions about the world and its role in that world. In the airborne, beliefs include the conviction that well-trained, well-armed troopers can fight out of any situation, for example. Beliefs and norms arise from the basic underlying assumptions an organization develops and maintains—unconscious and taken-for-granted ideas generated and refined through shared learning. These can be uncovered and deciphered by analyzing patterns in organizational practices. This book deciphers those values, beliefs, and norms from airborne units that manifested themselves throughout the larger army organizational structure during the early Cold War.[9]

Cultures are formed through the influence of organizational founders, the learning experiences of group members as the organization grows, and the introduction of values, beliefs, and assumptions by new members and leaders. Personality plays as much a role as anything. In the airborne, the actions of leaders in the early stages of unit formation contributed immeasurably to forming ideas about group behavior. What a leader prioritizes sends clear signals about what is essential within an organization. How leaders formalize ceremonies, awards, and rituals also indicates what a leader seeks to encourage an organization to value. Leader influence is evident in the airborne, as values developed by early officers—especially in doctrine—eventually permeated the entire airborne community. Once a culture takes root, it often defines for future generations what sort of leadership should be deemed acceptable by group members. Nevertheless, the actions of leaders in every generation represent key artifacts of organizational culture and continue to be powerful tools for communicating values to members.[10]

Building a new culture also involves a group's ability to "develop a shared concept of its ultimate survival problem" from its primary mission—its raison d'être. Culture often helps keep an organization together as it faces challenging situations. It represents a learned way of coping with stressors because the group learns a common framework of reference and interpretation for dealing with future challenges. Integrating new members into an organization is another challenge. Socialization through training, interaction, and observation is vital for communicating values, beliefs, and norms to new members. Messages transmitted to newcomers are essential to conveying the organization's basic assumptions and values.[11]

One of the natural evolutions of military history is to peel back the layers of broad military cultures to explore other cultures that lie beneath. Many historians have begun examining organizational culture, particularly how culture affects operations, including on the battlefield. Deciphering and understanding military cultures are paramount to understanding how and why battles turn the way they do, how wars are won or lost, the crafting of national strategy based on military capabilities, and resulting peacetime decisions military leaders make about institutional priorities and organizational change. Furthermore, examining the concept of a broad military culture represents an opportunity to study the amalgamation of subcultures within a large organizational structure. This begs for a more in-depth analysis of subcultures. It is unlikely for any sizable military organization to share a single set of norms and values. Services, branches, and even individual small units develop their ways of doing things based on their raison d'être that represent variations on the larger cultures in which they exist. These subcultures influence the larger culture as much as the larger military culture influences them in return.[12]

The airborne mafia do not represent the only large, selective organization in the military. The US Marine Corps offers an essential comparison for airborne organizations in the US Army. Culture is central to that service's narrative—especially beyond battlefield performance and into peacetime. As with the airborne, a nostalgic view of World War II was pivotal in how the US Marine Corps viewed itself and maintained relevancy in the Cold War. Like the role of airborne school in training paratroopers regardless of branch, every marine completes an identical boot camp focused on infantry skills. Like the marines, the airborne mafia used their shared experiences in World War II to shape their service's postwar organizational planning.[13]

The US Air Force's competing bomber and fighter factions provide further comparison. Malcolm Gladwell has thrust the concept of military mafias to the fore in his popular history, *The Bomber Mafia*. Like the strategic bombing advocates of Gladwell's story, the airborne mafia started as a small cadre of radical thinkers. Rather than believing in winning a war without ground forces, the airborne mafia insisted that the fusion of air and land power was critical to future warfare. Most works of military culture link cultural characteristics to wartime operational performance rather than look at how culture influences peacetime organizational changes. The Strategic Air Command (SAC) also offers an apt comparison, in which adherents to strategic bombardment doctrine grew to encompass a large subculture within the fledgling US Air Force and had an enormous role in that service's development during its first decade and a half of existence. Throughout its existence, the bomber mafia has competed with a fighter mafia; this fighter subculture developed alongside the bomber subculture, steeped in nostalgia and mythologies from World War I. The fighter mafia had an essential impact on developing and procuring equipment in peacetime service during the late Cold War. This volume demonstrates the ability of a subculture, like the fighter mafia, to affect massive changes in its parent service by harnessing peacetime nostalgia for wartime exploits.[14]

The adventures of paratroopers in World War II are well known, and this book does not try to re-create every detail of every employment of airborne forces during the war. Instead, it connects wartime experience to postwar organizational changes. In that postwar peacetime army, as its leaders began to look ahead to the next war, the impact of the airborne mafia is truly evident. As armor and airborne advocates wrestled for control of the service's identity and direction, both "claimed the mantle of the cavalry branch as the new warrior elite," according to historian Brian Linn. This phenomenon was best exemplified in the development of airmobile doctrine. Armor and air cavalry factions emphasized shock, protection, and firepower, while air cavalry could boast the ability to achieve surprise and mass on the enemy through vertical envelopment. Armor officers tended to dismiss airborne and light infantry forces as support for mechanized forces, while airborne officers did the reverse. Neither the armor nor airborne visions of future war was sufficient, as lightly armed paratroopers need protection in high-intensity conflict.[15]

In addition to describing the cultural origins of the airborne and how this group effected massive changes in the Cold War army, including the direction of national strategic decision-making, this book has important lessons for today's US Army. Thanks to the airborne mafia's influence on creating airmobile units, every modern infantry battalion is supposed to be capable of using helicopters. This study also demonstrates the capability of subcultures to dominate a larger institutional culture. While the airborne dominated the first half of the Cold War into the early years of the Vietnam War, the changing of the guard from Gen. William C. Westmoreland to Gen. Creighton Abrams as US Army chief of staff in 1972 (Bruce Palmer's one hundred days as service chief notwithstanding) represented a shift from an airborne- and infantry-dominated service to an armor-dominated service. Changes implemented in the post-Vietnam era to refocus army efforts on large-scale mechanized warfare in Europe represent that reality. Deterrence and planning to fight the Soviet army have constituted a significant portion of the army's intellectual effort since 1945. The renewed focus in the 1970s placed a premium on armored forces at the expense of lighter formations. To a lesser degree, the influence of airborne officers helps explain why the army maintained lighter forces throughout the Cold War while its primary potential adversary was mechanizing its forces.

CHAPTER 1

The Birth of American Airborne Culture

> The difficult we do immediately, the impossible takes a little longer.
>
> —Slogan on the wall of a hangar, Lawson Army Airfield, Fort Benning, Georgia

When Donald Deam returned from jump school in the summer of 1943, he expected to be in trouble with his boss. Deam was a young nonjumper and an orderly sergeant for the commander of the 501st Parachute Infantry Regiment (PIR), Col. Howard Johnson. Instead, Johnson promoted him to staff sergeant, exclaiming, "This is the kind of man I want in my outfit!" Deam was not supposed to attend airborne school, but as the colonel's orderly, he had sneaked his name onto the memorandum sending the next batch of trainees to Fort Benning. Deam gambled correctly that Johnson and the rest of the airborne required enterprising young men who wanted to be in these units badly enough to risk being thrown out. Deam became the first sergeant of the regimental headquarters company for much of the war. Johnson was killed in Holland while leading his regiment near the Waal River. Deam's experience, however, is indicative of a subculture that placed utmost emphasis on individual initiative and rewarded risk-taking.[1]

Attracting and training men who believed themselves and their units capable of "doing the impossible" was critical to creating a new unit type predicated on jumping from aircraft behind enemy lines. Pop culture images, such as the August 19, 1940, and May 12, 1941, *Life* magazine covers highlighted the new warrior ideal. The August 19 cover introduced Americans to the paratrooper, showing a fully deployed parachute canopy with parachutist hanging beneath. The May 12 cover depicted a stoic US Army parachutist named Hugh Randall awaiting his turn to jump; it inspired many Americans to volunteer for parachute duty. The article's eight-page spread of photographs and ominous language that "day after day, at the peril of their lives, brave men jumped" showcased the dangerous yet composed nature of the paratrooper. By September 1941, the army's propaganda machine helped produce the film *Parachute Battalion*, which had a similar effect. Parachuting was dangerous and cool, attracting those with a predilection for risk-taking. As a military technique, however, the concept was still in its infancy.[2]

Organizing units of men tasked with jumping out of airplanes was a new proposition in 1940. That unique raison d'être—parachuting—played a significant role in the development of the airborne subculture. An American made the first modern parachute jump in 1912, but the Soviet Union was the first to demonstrate large-scale airborne operations when it dropped two parachute battalions on an airfield outside Kyiv in front of an audience that included German, French, Italian, and US observers in 1935. Studying the Soviets, the Germans poured considerable resources into airborne development and conducted the first combat drop, sending airborne units into the Netherlands, Belgium, and Norway in 1940. On May 20, 1941, the Wehrmacht attempted to seize the island of Crete by air alone but suffered an initial casualty rate of 44 percent. Eleven days later, on May 31, the *Fallschirmjäger* dropped parachutists and gliders on four airfields and captured the island.[3]

Though the operation was recognized by British and American airborne theorists as a revolution in tactics, the high casualties signaled the end of major German airborne operations for the duration of the war. In subsequent campaigns, German parachute units were relegated to use as the Wehrmacht's corps d'élite—a mobile reserve of elite volunteer light infantry. Allied planners, however, saw in Crete the potential of entirely airborne armies. American theorists relied on the German and Soviet examples. Parachutist badge designer, early airborne pioneer, and future Green Beret advocate Capt. William P. Yarborough was dispatched to Moscow in late 1941 to study Red Army paratroop techniques, but he never made it because of the German offensive into the Soviet Union. James M. Gavin, also a captain at the time, "had access to many of the original documents relating to the German airborne operations in Holland" and reports from the invasion of Crete when he was writing the army's initial airborne doctrine.[4]

FIGURE 1. "Jump into the Fight" propaganda poster. The addition of parachute pay of fifty dollars a month more than doubled the 1942 monthly base pay of a private ($40) or a pfc. ($46). Image courtesy of the US Army Heritage and Education Center.

Nevertheless, the development of an airborne force in the United States was slow to materialize, compared to similar efforts in the Soviet and German militaries. The American experiment had begun on June 25, 1940, when US Army General Headquarters directed the Infantry School commandant at Fort Benning, Georgia, to seek volunteers for an airborne test platoon to begin jumping that August. The test platoon was created mere months after the successful German airborne operations of 1940. Expansion of the effort came quickly after that, and by September—fifteen months before America entered the war—army leaders formed the first parachute infantry battalion. Next came the airborne community's first command structure—the Provisional Parachute Group—established on March 10, 1941, to plan and execute training. Its first commander was Col. William C. Lee, who spent many months selling the idea to General Headquarters, United States Army. A larger headquarters, Airborne Command, was created in 1942 to meet the growing demands for a large force of paratroopers. The creation of Airborne Command put the fledgling paratrooper force on the same structural level as the traditional branches (infantry, cavalry, and artillery), reporting directly to Army Ground Forces. By the war's end, the United States fielded fifteen parachute infantry regiments, eleven glider infantry regiments, and three parachute infantry battalions that fought as part of, attached to, or independent from five airborne divisions.[5]

Developing Elitism

Like most elite units across history, the airborne relied on three primary components—voluntarism, special selection criteria and training, and distinctive clothing and insignia—to attract the individuals required to fill its ranks. Would-be paratroopers first had to volunteer for airborne training. Many airborne volunteers were originally drafted into the army yet chose to join the paratroops once in the service. Seventeen officers and over two hundred enlisted personnel volunteered for the one platoon leader and forty-eight paratrooper positions available in the original 1940 test platoon. In August 1941, four hundred enlisted men volunteered to join the newly formed 503rd Parachute Infantry Battalion despite requiring a reduction in rank to private. Second Lieutenant Richard "Dick" Winters chose parachute infantry over armor because "they looked impressive, were physically fit, and demonstrated what I could only call a tolerant scorn for any soldier who was not airborne qualified. I wanted to be with the best, and paratroopers were the cream of the crop. I volunteered immediately to become a paratrooper."[6]

Hand-selected recruiters traversed bases in the United States looking for volunteers who embodied the airborne ideal. Commanders designated men for this duty based on their fit appearance and ability to recruit like-minded volunteers from the various training bases that dotted the American landscape. Pfc. Vincent Speranza—a replacement in the 501st PIR after Operation Market Garden—described the airborne recruiters he saw as "magnificent men in sharp uniforms, brilliantly shined boots, and glittering silver wings on their chest." Money was also an important incentive. Billy Pettit of the 511th PIR—a man who had never seen an airplane before—commented that when he inquired about the airborne, the sergeant replied, "I don't know much about this paratrooper business, but they get $50 jump pay." That was enough for Pettit to join. Many volunteers were attracted to the daredevil nature of the occupation and its distinctive symbols—a form of self-selection in which like-minded individuals already prone to dangerous behavior chose to join the paratroops. Money, prestige, testing oneself, and the prospect of excitement were significant motivating factors in finding the personnel required to fill parachute units. Young men struggling to live up to the new idealized wartime male image were enamored with the prospect of jumping from airplanes—even those who had never seen an airplane before.[7]

The training was too danger-prone for the army to assign men to parachute duty haphazardly—volunteers were required. This voluntarism became a fundamental distinguishing feature of the airborne regiments in an army of draftees. Just to volunteer, prospective paratroopers had to meet specific criteria. In addition to strict physical standards, paratroopers also had to score 110 or higher on the Army General Classification Test (AGCT), the same score required for those wishing to attend Officer Candidate School or enter the Army Air Forces. This test was designed to measure an inductee's intelligence and determine the type of military occupational specialty an inductee was suitable for. Numerical scores were grouped into five classes. Class 1 represented the men of highest intelligence and Class 5 the lowest. All arms and services were supposed to receive a proportionate distribution of men from all five classes. While the rest of the army struggled with finding capable recruits, the airborne received preferential treatment because of "the special intricacy of their problems." The army allowed airborne divisions to dismiss any soldiers who scored in Classes 4 and 5 on the

AGCT, which gave airborne units a much higher percentage of Class 1 men than the average ground forces divisions. This policy allowed the airborne to keep only the smartest and best-qualified men and ensured special treatment in manpower policies.[8]

Qualifying to volunteer was only the beginning of separating the potential paratrooper from the ordinary soldier. Pvt. Kurt Gabel of the 513th PIR explained, "In my euphoria at having passed the magic score of 110, the first barrier to the exclusive club of the paratroops, I was in a little world of my own and lost what slight identification I had with my fellow draftees." In addition to AGCT scores, prospective paratroops then had to meet stringent physical requirements: age between 21 and 32, no more than 185 pounds, and minimum visual acuity of 20/40 in each eye. Selections of volunteers were to be further screened at the unit level for "demonstrated soldierly qualities, agility, athletic ability, intelligence, initiative, determination, and daring." These standards were higher than those for regular recruits—this was the next rung on the ladder separating these volunteers from the rest of the army. Paratroopers believed they were better because they met the standards for *volunteering*. The sheer number of volunteers enabled commanders to select only the most intelligent and fit men to form their units.[9]

Because the parachute troops were all volunteers, and commanders had the final say in staffing their units, voluntarism became another gate of entry. The initial doctrinal manual for airborne forces, Field Manual 31-30, directed that "where choice of volunteer personnel is possible, preference should be given to those of an active, agile type." The 511th PIR, for example, included a timed run up Currahee Mountain (elevation 1,735 feet) at Camp Toccoa in Georgia to gauge hardiness for training. The regiment wanted fit, intelligent, and audacious men who could make split-second decisions, and it had the latitude to find them. Of 12,000 original volunteers, the 511th brought 2,176 from its basic training at Camp Toccoa to the Parachute School at Fort Benning. The experience of crafting the 506th PIR is likewise typical—5,800 volunteers for 2,000 positions. The 506th, according to William Guarnere, was whittled down to 1,500 enlisted men before they went to jump school, mostly owing to an inability to keep up with physical training. Underlying the entire course was basic psychology to dispel the fear of heights and translate that confidence into combat. Dick Winters understood how the training steeled his paratroopers' hearts against the tragedy of death. "I don't believe that, as paratroopers, we faced the shock of our first fatality to the degree that most outfits do in combat.

Every paratrooper encounters the possibility of serious injury or death on every jump." Physical exercise was for far more than just building muscles; it was used to build automatic responses—of the sort required to keep men alive on the drop zone and in combat.[10]

New paratroop recruits encountered psychological screening on arrival. The cadre asked volunteers to jump from the notorious thirty-four-foot training tower, dismissing any hesitant volunteer. The same was true for any exhibition of fear throughout their training. Jumping required soldiers to control their fears to accomplish their mission—just like in combat. Gabel described further opportunities to quit before the training intensified, and many did. For those regiments that began at Fort Benning rather than Camp Toccoa, "The Frying Pan" represented another selection process that would-be paratroopers experienced before assignment to their regiments. The Frying Pan was a training area so named for the hours of physical toil recruits spent there under the hot Georgia sun. After surviving his Frying Pan experience, Gabel and the 513th PIR then spent thirteen weeks in collective unit basic training before attending the Parachute School, a similar process to that of the regiments formed at Camp Toccoa.[11]

Perhaps just as important as the training itself was the belief it was more challenging than that endured by other soldiers. Col. Mark Alexander, who commanded three separate battalions during the war, called jump school "some of the toughest physical training I have ever experienced." The belief that one's endurance was superior fostered an elitist attitude; whether it was elite or not was irrelevant. Donald Burgett remembered that during one of the morning runs "we kept going on and on without any sign of the break we were used to getting in the regular infantry." Maj. Gen. Matthew B. Ridgway—commander of the 82nd Airborne Division—believed their training developed the men's fighting spirit while conditioning them to championship boxer level. Its rigorous nature, high failure rates, and emphasis on physical fitness and mental toughness created cohesive fighting units that viewed themselves as different and superior to the rest of the army. The men who made it through selection and endured training believed this, as did their commanders.[12]

Ridgway found his way to the 82nd by assignment rather than volunteering. He had served alongside George Marshall in the War Plans Division in Washington, preparing for war with Japan or Germany throughout 1941. After Pearl Harbor, he waited patiently for word from Marshall about whether he would be given a troop command. At the time, he was a lieutenant colonel, but by January 1942 he was advanced to brigadier general and assigned as the assistant division commander for the 82nd Infantry Division under Maj. Gen. Omar Bradley. Bradley was reassigned to the 28th Infantry Division in June 1942, leaving Ridgway in charge. By August of that year he wore the two stars of a major general and was ordered to convert the unit into an airborne division. "My knowledge of airborne operations at that time was exactly nil," Ridgway recalled, but he set about learning and turning his unit into a first-rate outfit. His leadership style—that he would never order his men to do anything he was unwilling to do himself—was well suited to help shape airborne culture.[13]

After completing their basic training, new paratroopers entered the Parachute School at Fort Benning. The Parachute School consisted of four lettered stages, A through D. Physical conditioning comprised most of stage A, but units that arrived at the Parachute School together generally skipped this requirement because their four-month-long basic training hardened their bodies more than a week at Fort Benning ever could. Stage B consisted of learning the basic techniques of how to tumble upon landing and how to manipulate the parachute risers to slip away from fellow jumpers or obstacles. The students then practiced the proper jump procedures, first from an airplane mockup and then the infamous thirty-four-foot tower. Stage C included more training from the thirty-four-foot tower but centered on using the 250-foot "free towers" installed at the center of Fort Benning. Inspired by similar towers constructed for the 1939 World's Fair in New York City, the 250-foot towers allowed the prospective paratrooper to practice controlling his descent and make a proper parachute landing fall, all while receiving coaching from instructors. Stage D consisted of five jumps from a C-47 aircraft—the fourth jump introduced combat equipment, and the fifth and final jump was a nighttime operation with assembly by unit, as if in combat. After completing Stage D, the young paratroopers were awarded their coveted jump wings.[14]

The American airborne forces of World War II were known for distinctive uniform items. Jump boots, the baggy pants of their jumpsuits, jump wings, unit flashes behind those wings, and parachute patches on their garrison caps all designated the wearer as something special. Authorization for blousing their pants into their jump boots in dress uniform came in January 1941, followed by the parachute patch for wear on the overseas garrison cap. The silver parachutist wings, still awarded to Parachute School graduates to this day, were designed by test platoon member 1st Lt. William P. Yarborough in March 1941 and awarded a patent in 1942. Babe Heffron commented on the power of the jump wings when he noted that "your jump wings mean more than anything. You can have your Purple Heart or Bronze Star, but don't ever take the jump wings." The wings carried an almost mystical quality. Cargo-pocketed baggy pants tucked into high leather boots were borrowed from the German *Fallschirmjäger* uniform debuted in their 1940 operations in Norway and the Low Countries. Loose pants appeared in other elite units throughout history, such as French Zouaves and their predecessors, the Ottoman Janissaries. Looser pants can allow increased comfort and agility when moving; deep pockets provide more places to stow equipment on the jump or on the move. Like the airborne, Zouaves and Janissaries also enjoyed special status and prestige; all wore baggy pants to designate their differences. Unique uniforms—especially baggy pants—are integral to elite, specialized units throughout modern history.[15]

A distinctive two-piece jumpsuit, also designed by Yarborough, replaced earlier mechanics' coveralls, and an earlier two-piece suit became the field uniform of choice in 1942. The uniform continued to evolve; cargo pockets were enhanced with reinforced stitching after the first combat jumps in 1942 and 1943. Yarborough also helped design the jump boots and create many airborne uniform elements, including oval-shaped wing backgrounds bearing regimental colors. At a time when regular infantrymen wore canvas gaiters around their lower legs that gave them a straight-leg appearance, paratroopers' bloused boots created a distinctive look. A wartime survey from sociologist Samuel Stouffer showed that jump boots were the most iconic symbol of paratroopers. Stouffer's team found in 1943 that "seventy-five percent of 500 interviewees noted the jump boot as the most distinguishing mark of the paratrooper. Ninety-five percent said the jump boots meant a great deal to them, and three-quarters who did not have them claimed they would pay over $16 per pair (approximately 1/3 of parachute pay) out of their pocket if they could." The uniform, boots, and wings cemented the paratroopers' image as different from ordinary soldiers—a critical component of elitism. The jump uniform, jump boots, and parachutist badge were tangible items that showed the rest of the army that the wearer was special. Maj. William "Bud" Miley—commander of the first parachute battalion—authorized each of these characteristic, recognizable group identifiers and awarded them to each paratrooper as he completed parachute training.[16]

Francis Sampson, regimental chaplain for the 501st PIR, described the transformation into arrogant, egotistical paratroopers after parachute training: "We then hastened to acquire the overbearing mannerisms and obnoxious characteristics of pre-combat paratroopers." As could be expected, cocksure paratroopers caused problems at the various training bases and towns in which they found themselves around the country. In places like Columbus, Georgia, and Fayetteville, North Carolina, fights broke out between paratroopers and any of the myriad nonparatroopers around the base. Anytime a military policeman arrested a paratrooper for being drunk, out of uniform, fighting, or destroying private property in an off-post bar, more brawling could ensue. Even the army's highest-level commanders noticed arrogance emanating from parachute regiments. At a Washington press conference on paratroopers, Lt. Gen. Lesley J. McNair, commander of Army Ground Forces, remarked, "They [the paratroopers] . . . are our *problem children*. They make a lot of money, and they know they're good. This makes them a little temperamental, but they're great soldiers." Gavin agreed it was better to learn to live with the problem and "tolerate a certain amount of their misbehavior but have guys who were really capable fighters and confident and proud." Putting up with boisterous behavior was one risk associated with creating units that believed themselves better than anyone else.[17]

Drinking beer and singing macabre songs were some of the paratrooper's favorite pastimes. Bob Bearden of the 507th Parachute Infantry marveled at how his instructors drank all night in Columbus or nearby Phenix City, Alabama. They still led recruits on early morning five-mile runs, "barking commands all the way." A peculiar, ritualistic tradition known as the prop blast served as another practice for building cohesion among the units' officer corps. The 501st Parachute Infantry Battalion, commanded by Miley, started the tradition by holding the first ceremony. Beginning with the battalion commander and then in descending rank order, each blastee drank a concoction of vodka, lemon juice, sugar, and champagne out of a 75mm casing with two ripcord handles welded on either side, which became known as the Miley Mug. Each officer climbed onto a chair, jumped, and attempted to perform a satisfactory parachute landing fall. The blastee then jumped to his feet and downed the contents of the mug while the other officers counted "one-thousand, two-thousand, three-thousand" as on a jump. The blastee was supposed to finish during the count or face a penalty of more drink. The dangerous nature of parachuting encouraged other forms of cynical humor, inspiring memorable ballads such as "Beautiful Streamer," set to the tune of Bing Cosby's 1940 rendition of Stephen Foster's "Beautiful Dreamer," and the perennial favorite, "Blood on the Risers," set to the tune of "Battle Hymn of the Republic."[18]

Even leaving a parachute unit was a ritualized experience. In the 506th PIR, each soon-to-be-former unit member had his patches ripped off and was forced to un-blouse his boots and change into regular infantry shoes before marching off. A drummer maintained a slow, ominous roll with his drumsticks and snare drum throughout this morose ceremony. The result of this ritual was unmistakable—the regiment experienced fewer problems with discipline, particularly late returnees from furlough. Other regiments simply insinuated that tardiness meant removal from the paratroops. Being a member of the airborne was so revered and held in such high esteem that members of this fraternity were reluctant to leave their unit at any cost—especially if that meant letting down the cohesive teams they had spent months building.[19]

The Airborne Command's emphasis on voluntarism, special selection and training, and distinctive insignia served to create a highly motivated group with confidence in self and collective efficacy. Ridgway understood the importance of molding units during this time, especially the power of symbols and unit pride: "Little things can help greatly, guidons [unit flags], signs in your areas, letterheads on your official papers," and other ephemera stressed to individuals their place in the larger organization, thus deepening pride in their unit. Yet simply believing in their superiority was as essential as any other facet of the paratrooper mystique. This belief produced an enhanced sense of importance, that these men and their unit could handle anything that came their way. For a group that would soon have to jump into enemy-held territory and regard isolation as a normal battlefield condition, believing in their collective elitism was critical to mission success.[20]

Flexible, Innovative, and Adaptive Leadership

Airborne operations required immense flexibility and improvisation, often more than other military operations. The ability to innovate and respond to changing conditions was paramount to success. Furthermore, this tenet, alongside the inherent lightness of airborne divisions (fewer men and less equipment, firepower, and transportation), helped breed an expeditionary mindset comfortable responding to various contingencies. Leaders stressed to their men that they would endure more significant hardships than regular infantrymen and therefore had to excel at everything asked of them. Flexibility became an integral part of the airborne's capabilities. Paratroopers were supposed to meet the enemy in any direction immediately upon landing; the vital need for resilience and adaptability meant that every unit, from the squad to regiment, needed to be as self-sufficient as possible.[21]

Division structures were thus organized to achieve maximum flexibility. The airborne division was never considered a true division but rather a headquarters structure to command regiments and other forces as needed. Deliberate training for flexibility emphasized capabilities needed on the drop zone that were often outside the regular infantryman's scope. While airborne divisions were designed lighter than regular infantry divisions, the creation of actual divisions did not occur until 1942, as most commanders viewed parachute units as enhanced commandos with parachute capability. A parachute infantry platoon originally consisted of two maneuver squads, unlike the three in a regular infantry platoon. Glider infantry regiments had only two battalions, compared to the three in all other infantry regiments—including parachute regiments. Fire support and transportation assets were likewise far fewer. The division consisted of 8,600 personnel, compared to the roughly 14,000 in a standard infantry division. This overall streamlining effect forced commanders to consider how and when to apply their forces and for how long, even during training. With so few personnel, sustained combat operations were impossible, especially in unit structures requiring specially trained volunteers.[22]

Training inculcated flexibility to prepare for battlefield conditions, and shortages in training equipment further reinforced this. A significant factor compounding training problems came from the Army Air Forces' inability to provide adequate transport aircraft for training. Lend-Lease and the burgeoning airplane industry pulled pilots and airframes to all corners of the globe, leaving aircraft and pilots for parachute training in short supply. The independent-minded Air Forces also preferred to focus efforts on equipment that might prove their worth as a separate service, requisitioning larger numbers of bombers and fighters—pilots and aircraft—than less-exciting transport planes. The shortage of planes hamstrung the growth of a new branch whose very existence depended on parachuting from flying aircraft. Aircraft shortages continued well into the airborne concept's development as the 501st Parachute Battalion trained with only twelve aircraft through the entirety of 1941. Each C-47 carried a maximum of twenty-eight paratroopers, so there was not enough capacity for the entire battalion. The largest drops could carry two companies with attached enablers—if there were no aircraft maintenance issues. According to the Combined Chiefs of Staff in April 1942, the army had only seventeen aircraft available for airborne training and made plans to requisition all civilian airplanes in the United States to carry out future airborne operations in Europe.[23]

Under these circumstances, the Army Ground Forces commander, Gen. Lesley J. McNair, did not see the need for more than a handful of mass tactical airborne operations in training before sending units overseas. Problematic combat jumps in Sicily and Italy revealed the error in McNair's thinking. Shortages in the number of gliders available for training were even more problematic and played a significant role in reorganizing airborne divisions' structure. Indicative of a country shifting from a peacetime to a wartime economy, shortages extended throughout the fledgling airborne experiment, from parachutes to radios, living quarters to training facilities, and even ammunition. The need for equipment became so acute that the original test platoon and subsequent classes wore modified football helmets and practiced landings by jumping out of the back of troop trucks while they drove fifteen miles per hour down dirt roads. Proper parachutist training equipment did not yet exist, and commanders learned to make do with what they had.[24]

Despite the lack of resources and equipment, airborne soldiers needed to continue to train and develop flexibility to handle their potential battlefield conditions. To help meet this goal, training plans included increased attention to weapons proficiency across the platoon and company levels. In the regular infantry, crew-served weapons such as machine guns and mortars were used by designated personnel. In the paratroops, every man was required to understand and operate those complex weapons to offset the impact of casualties and dispersal. The first official doctrine concerning airborne operations, Field Manual 31-30, emphasized the difference between training regular and parachute infantry units of all levels: "Unit training of parachute troops closely approximates that of rifle regiment of comparable size. It differs principally in that all parachutists must be qualified to handle all platoon weapons. . . . In addition, parachute troops must be trained in executing demolitions." Gavin noted that the "problems were without precedent. Individuals had to be capable of fighting at once against any opposition they met on landing. . . . We had to train our individuals to fight for hours and days, if necessary, without being part of a formal organization." This attitude emphasized empowering subordinates who understood they must solve each problem at the lowest echelon. These men were highly qualified infantry prepared to fight with minimum support while surrounded by enemy forces. This mindset undergirded every aspect of airborne training.[25]

Creating flexible paratroopers was the natural byproduct of tough and realistic preparations. Hands-on field training followed every classroom block of instruction as the training cadre accounted for every minute; there was no time to spare in preparing these men for combat. In a training memorandum from Airborne Command to the 82nd and 101st Airborne Divisions, the higher headquarters stipulated that "all individuals will be conditioned to withstand extreme fatigue, loss of sleep, limited rations, and existence in the field with only the equipment that can be carried by parachute, glider, or transport aircraft." Gabel registered these techniques' effects, noting that their experiences made his unit feel that they "were already veterans" by the time they entered combat. Spencer Wurst of the 507th PIR remarked that his unit "did more shooting on tactical problems than the regular infantry." He also commented that they concentrated more on difficult night training than other infantry units.[26]

The same memorandum from Airborne Command also reminded airborne leaders that "all units will be prepared to either enter combat immediately on landing or move promptly by marching against an objective." Donald Burgett, a machine gunner in the 506th PIR, noted the cumulative effect of his training and its role in preparing him to meet the enemy in combat: "The rules were simple: operate with automatic efficiency as the result of a deliberate process of manipulation." Commanders and planners knew ahead of time that flexibility on the battlefield was necessary for successful units and acted with this in mind through every step of their training regimens. Flexibility was integral to the airborne experiment from its inception in 1940 throughout the war and beyond. Maintaining a lighter unit structure in a unit that stressed versatility, adaptability, and air transportability bred an expeditionary ethos that ensured airborne units were prepared to respond to crises.[27]

Improvisation, innovation, and an ability to adapt to changing situations were hallmarks of the airborne mindset during World War II. Pioneering a novel way of warfare attracted innovative minds, the same sort of minds that would find ways to transform the military wherever they went. Leaders who exhibited a penchant for new ideas in a parochial and static interwar army—one still teaching horsemanship to officers at the onset of war—found a home in Airborne Command. As the head of the unit, William C. Lee encouraged innovation. Small-group tactics and the potential for scattered drops required troopers at all echelons, from squad to division, who were comfortable improvising and adapting to the situation at hand—often forming ad hoc small units. This ability to fight in what became known as LGOPs—little groups of paratroopers—became a trademark of airborne soldiers throughout the war and continues to undergird airborne culture into the twenty-first century.[28]

In 1941, James Gavin, a captain teaching at West Point, had difficulty persuading the academy to release him for airborne training, so he wrote to the test platoon leader William T. Ryder for help. Ryder's response emphasized the need for progressive, innovative officers in the airborne. "I know that you will like the [parachute] service, because there is nothing to this jumping out of an airplane, and there is a great need for ideas," Ryder wrote. "We are still suffering from growing pains, and if there was ever a place in the Army where people were willing to listen to your ideas—this is it! The older officers are always willing to listen to the jumping officers' point of view." Ryder helped bring Gavin to the paratroops, persuading Lee to ask the army to override West Point's decision. That young officers had the forum to express opinions and test new ideas represented a shift in mindset for a hierarchical organization like the US Army. Perhaps no one exhibited more innovative savvy than Gavin, who upon being assigned as the Provisional Parachute Group S-3 operations officer proceeded to write the first manual for airborne operations, Field Manual 31–30, *Tactics and Technique of Air-borne Troops*, based on his knowledge of Soviet and German doctrine.[29]

Maxwell Taylor, in contrast, was not an enthusiastic paratrooper. Yet he found himself the beneficiary of working for Gen. George Marshall in 1942. After Joseph Stilwell requested his services in the China-Burma-India theater (Taylor spoke fluent Japanese), Marshall approved Ridgway's request for Taylor to serve as *his* division chief of staff instead. Taylor found most paratroop officers overly enamored with jumping and too quick to excuse rowdy, undisciplined behavior. Rather than the hundreds of jumps logged by other officers, he made no more than six in his entire career, two of which were combat jumps. "I viewed the parachute strictly as a vehicle to ride to the battle, to be used only when a better ride was not available." In this regard, Taylor was an opportunist who was requested by name and made the most of his service in the airborne.[30]

Leaders willing to share every hardship with their men are often an elusive ideal. Stories abound of poor leaders throughout all units, including the airborne. Nevertheless, leading from the front was a requirement, as paratroop officers always led their men out of the jump door; standard operating procedures dictated that officers jump first. Commanders used this ideal to promote and showcase the leadership necessary for bolstering the leader's image and capacity to maintain control and respect under fire. Furthermore, because they needed to fight in a dispersed manner after landing, airborne units required "prompt, decisive, and intelligent leadership," as outlined in the field manual. More so than anywhere else, leaders in the airborne were required to embody the ideal they expected of their men. Airborne officers were always with their men and sharing their hardships. Of course, this is not unique to airborne units; examples of dynamic leadership abound in regular combat arms formations. The difference is that airborne regiments sought to attract men who exhibited these traits, while also cultivating dynamic and inspired leadership from within the ranks.[31]

Having leaders who had earned their men's respect was critical to success in the sort of desperate situations paratroops might find themselves. "They were leaders from the general on down," Pvt. Irvin Seelye of the 505th PIR recalled decades later. Taylor tells the story of a chaplain assigned to the 82nd Airborne Division who had problems understanding his unique flock. The chaplain all but begged Taylor for a chance to prove himself by attending the Parachute School. Taylor relented. "I let him go to parachute school and he came back a changed man. Henceforth, he was one of the boys, and they accepted him as such. . . . Throughout the war, he was a tower of spiritual strength among the men." This sort of front-line leadership was crucial to engender mutual trust within individual units. Samuel Stouffer's survey found that the best, most-respected officers were those who were "always with [their] men in combat and led by personal example," yet the researchers also found that 75 percent of enlisted respondents agreed "that most officers are more interested in getting promoted than in doing a good job." To be sure, many airborne officers also suffered from this lure, though airborne units generally proved the exception. The division commander and a private first class might find themselves alone on a drop zone in the middle of the night, and it is for this reason that everyone had to meet the same standard, and why leaders led with conviction. This trained-for tenet separated the airborne from the rest of the infantry.[32]

The press noticed, too, and reported that paratroopers possessed a close-knit identity forged in a communal crucible unique to their profession. "Everybody in a parachute outfit jumps, the chaplain, the doctors, the adjutant and the cooks and the K.P.'s and the clerks," the journalist Marshall Andrews reported. The reason for this was simple: if everyone was not a parachutist, "a hard-boiled paratrooper might not show the respect and confidence a doctor or a chaplain deserves." The *Chicago Tribune* correspondent Jack Thompson noticed as well: "The morale of these young troops is excellent. One of the reasons for this is the close association of officers and men. All get identical training. They jump together and every soldier knows that whatever he has been asked to do, every officer can do. They share the same rations and, in the field, live under identical conditions." Even the public relations officers trained like infantrymen—every man was supposed to jump into combat.[33]

Owing to the heightened sense of occupational danger, leaders understood that unit cohesion and control hinged on reinforcing their capability to endure whatever their men endured. While preparing to become the 82nd Airborne Division's chief of staff, Taylor realized that to control this rowdy bunch of paratroopers who thought they were better than everyone else meant that he and the rest of their new leaders needed to jump, too. "It was apparent to us visitors that if we were to impose our authority and the discipline needed on these troops," Taylor wrote in his memoir, "the first thing to do was to get a parachute jump to our credit." Officers also shared the load on long marches, often carrying machine guns or mortar tubes; the men's welfare became their top priority, and by and large airborne leaders demonstrated this to their units. Paratroopers knew they could trust their officers. Even during parachute training, "we had the officers jump out first," Andrew Carrico of the 511th PIR recalled, "so, it is up to the officer to show you weren't afraid." Sometimes, to reinforce the importance of leadership by example, the platoon leader was relieved alongside his soldier if a trooper failed to jump.[34]

Airborne units carried the expectation of independent operations far from higher headquarters, in a mindset that persisted from division to squad level. Trust in leadership was critical for building and maintaining esprit de corps for the airborne. Leaders knew they had to "be quite willing to do anything you asked them to do" and do it better than the men. Leaders stressed the need for innovative personnel flexible enough to train despite equipment shortages; they also recognized the need to balance cohesiveness with individual thinkers prepared for decentralized operations. Recent psychological studies have vindicated the efficacy of this kind of training in developing leadership traits. Identification with and faith in the larger organization are essential for what sociologists call vertical bonding or hierarchical cohesion. In the military, this concept equates to pride in one's larger organization above the primary group, and soldiers often identify with multiple levels, such as their battalion or regiment. In World War II, however, divisional loyalty was common in the US Army owing to identifiable division patches on each soldier's left shoulder. This concept was enhanced in airborne units as a result of the unique elitism bred within the organization, the distinctive uniform items, and trust in leaders up and down the chain of command.[35]

Individualism, Decentralization, and Cohesion

Operating in small units far from friendly support posed a real challenge to paratroopers in World War II. Jumping behind enemy positions, with the subsequent need for small units and individuals to operate independently, was new for an army still mired in close-order drill. During the beginning of the airborne project, leaders realized a need to break from the norm, taking measures to stress the individual paratrooper's importance. The reality of airborne operations practically guaranteed that even the most accurate drop would scatter troopers over a wide area, forcing small groups to fight for days as independent, ad hoc units. Therefore, Airborne Command stressed its units' ability to operate in a decentralized fashion, with a doctrinal requirement to "be prepared to either enter combat immediately on landing or move promptly by marching against an objective." To prepare for this eventuality, leaders gambled that emphasizing the individual would enhance the group's overall efficacy, which paid off with immense success in combat.[36]

Because every member of a parachute unit, from commander to private, might find himself in the same situation—scattered and with only small groups forming a command—following their jump, every member of the unit needed to think of himself as a self-sufficient actor on the battlefield. All paratroopers needed to understand the mission and their role in achieving their objectives. Young officers in airborne units thus promoted individuality and found success when leading isolated small units—sometimes without formal leadership—and often with men from adjacent units. The initial field manual stressed such an eventuality, stating that every member of the parachute force "must seek decisive action immediately upon landing. Success depends largely upon rapid execution of missions assigned to subordinate units," and "failure of one of the smaller units to accomplish its mission may mean defeat of the entire parachute command involved." This reality represented a unique problem; while most infantry units arrive at the battlefield intact and disperse when the fighting commences, parachute units enter the battlefield even more dispersed and had to come together under fire.[37]

In his memoir, Gavin elaborates on how airborne leaders sought to emphasize individualism. He recognized the unique nature of the issue, writing that all the problems he and his fellow officers were trying to solve "brought into sharp focus the most important problem of all—how to train the individual paratrooper." Gavin and his contemporaries understood that paratroopers sometimes had to fight on their own for days (or even weeks) and thus needed a strong, independent mentality. As one of Gavin's subordinate battalion commanders later wrote, "The last thing he wanted was a regiment of automatons who did not act until told to do so. This applied from the top all the way down to the lowest private." In a relatively small but important move, Gavin and his fellow airborne leaders emphasized self-sufficiency by pioneering name tags on the uniform—something antithetical to most military units' standard operating procedures that stressed uniformity. Previously, armies had relied on close-order drill to build and maintain cohesion. This is also not to say that drills and ceremonies no longer had a place in military training—even in the airborne. Marching and running together, in formation, continued to serve as building blocks for training the whole and proved essential to creating cohesive units later.[38]

Gavin's ideas surrounding individualism had a threefold impact. First, they stressed airborne distinction—regular infantry had not yet adopted name tags. Second, they taught individual paratroopers that they were superior and capable of anything asked of them. Third, Gavin reinforced the need for soldiers on the battlefield to think for themselves. Leaders encouraged paratroopers to articulate their tactical opinions and try new ideas. Critical thinking about leadership and understanding the mission helped junior leaders take charge when senior leaders were absent or became casualties. For Gavin, this was one of the great successes of stressing individuality. In his memoir, he wrote that "there were many occasions in combat when the paratroopers were mixed up with regular ground formations, and paratrooper officers and NCOs effectively took over the command of larger formations of other units. Aside from the impact of this type of training on airborne formations themselves, it had a tremendous significance to the army as a whole."[39]

The balance between individualism and discipline has been a routine problem in Western militaries since the creation of light infantry in the seventeenth century. The French revolutionary-era Armée du Nord of the late eighteenth century emphasized the individual in developing *tirailleurs* (skirmishers). Achieving compliance in individual soldiers without close supervision and coercion became possible thanks to French nationalism and a belief in the cause. The *tirailleur*, in turn, relied on the confidence he had in his comrades—something that could only develop through time and experience. This trend continued in the United States as well. By the end of the nineteenth century, American trainers placed increasing emphasis on skirmish formations. During the 1880s and 1890s, the US Army abandoned stifling linear tactics and tedious drills to enhance the individuality of junior soldiers and leadership of noncommissioned and junior officers in the field. The German *Stroßtruppen* of the First World War represents another evolution in the decentralization of infantry combat and reliance on junior leaders and individual squad movement. Tasked as they were with infiltration behind enemy lines as rapidly as possible, reliance on a keen understanding of the mission and the independence to act within that intent prioritized decision-making at the lowest tactical level, which was necessary to ensure their ability to exploit an advantage gained.[40] Since the advent of firearms, the infantryman has been undergoing an evolution of tactical individualism on the battlefield. By World War II, all infantry could fight as light infantry, and the airborne represented the apotheosis of this centuries-long trend. Airborne forces emphasized individuality, dispersed operations, a light equipment footprint, weapons expertise, junior leadership, and athleticism. The primary difference was how airborne units emphasized these traits to mitigate the unique battlefield condition of starting dispersed resulting from arrival by parachute behind the enemy's lines.[41]

When faced with an entrenched enemy, taking the initiative and carrying it through to an objective could help smaller units maintain a psychological advantage over their adversaries. One way to train that principle included rotating people in leadership positions during exercises. In contrast to those in airborne unit training, many regular infantrymen lacked initiative or aggressiveness in combat, reflecting bureaucratized training regimens that emphasized overbearing leadership. In the 506th PIR, Bill Guarnere noted, "the kids took turns being an officer or a sergeant out in the field—they tried to prepare everyone to take over if you had to." Of course, this was a reality in any combat unit, but parachute units trained for that eventuality. Because units rarely realized who was a casualty until days into the operation, taking charge of larger units was a critical capability practiced in airborne units.[42]

Airborne leaders, of course, could not focus purely on battlefield autonomy and individualism. Squad-size training and small-unit tactics were another trademark of the training process, also steeped in the reality of potential dispersal behind enemy lines. Recognizing the importance of small units for overall mission success, Field Manual 31-30 stated, "When operating as part of a larger unit, the missions of the parachute squad are substantially the same as those for any other infantry squad. However, the parachute squad usually is given much greater freedom of action than the infantry squad and may act independently." To meet these requirements, battalion commanders such as Richard Seitz of 2nd Battalion, 517th PIR, stressed that "the attention to squad and section training was considered important because when the squad or jump stick hit the ground after the jump, they would be the first and only group with immediate tactical or unit integrity. They would be required to fight independently during the initial phase of action. All parachute units accepted this doctrine." Squad-level training is the benchmark for any combat organization. Still, it is essential for airborne units that will find themselves in ad hoc squads made of members from across their regiment or divisions.[43]

The training schedule issued to parachute infantry battalions accordingly emphasized small units. After thirteen weeks of basic training, fourteen weeks were dedicated to squad, platoon, and company training, while the battalion training phase lasted two weeks. (Of course, wartime constraints played a vital role in this truncated training schedule.) Henry Langrehr recalled the necessary emphasis: "So much of my training, and [that of] the rest of the units, centered around what we called small group tactics in the army. We learned to engage the enemy in multiple ways, usually without support. . . . I was expected to be able to do it for days and even weeks, if necessary, without an officer or even a sergeant directing me. Heady stuff for a nineteen-year-old."[44]

These ideas were unorthodox in the army of the early 1940s. Army-wide efforts focused on battalion-and-above collective training in preparing infantry divisions for combat. Infantry divisions trained from squad to regiment over eleven weeks, with a second eleven-week period focused on combined-arms training above the regimental level—a full five fewer weeks of unit training. The most important factor in training (and combat) was leadership, and most divisions at least reached an acceptable level of proficiency despite less focus on small-unit training than airborne units. Those leaders who emphasized training below the battalion level found their units more effective than those who did not. Airborne leaders understood that they were asking their paratroopers to exhibit abilities usually reserved for commandos and special-mission units trained to perform reconnaissance or demolition raids. Leaders expected their men to succeed in the most desperate situations by stressing the individual's capability. The division yearbook noted, "There can be no such animal as a typical parachutist. Every 82nd Airborne trooper is by the nature of his mode of warfare an individualist of the first rank."[45] This training regimen instilled in each trooper the belief that he would succeed in dire circumstances. It might seem antithetical that training individualism might foster cohesion, but that was precisely the net effect. From the beginning of their training, paratroopers understood that they were volunteering for something hazardous. The skills and confidence developed from proving themselves worthy of inclusion were influential in fostering collective confidence in their and their unit's ability to survive and even succeed in combat.

Military skills like shooting and maneuvering were essential, but the relationships developed and cohesion engendered in their training yielded strong primary group ties among airborne soldiers as they completed what they believed was tougher schooling than any other army unit. And as Gabel remembered, "morale and motivation continued to grow as our confidence increased." Motivation, morale, and confidence grew from knowing the men with whom they were soon to be fighting alongside. "We were a family, way before we hit the battlefield," remembered Guarnere. Richard Eaton, a member of the 517th PIR, remarked that the isolation of training "in the hinterlands of Georgia and North Carolina undoubtedly contributed to the esprit that developed." Even regimental and battalion staffs stressed group cohesion. And confronting the shared danger of jumping imbued these units with cohesion, which played an instrumental role in setting them apart from regular units.[46]

Cohesion in airborne units was built through shared hardship and performance. Members of units unite during training toward a shared common goal. This execution of collective practices cemented bonds that felt very much like personal affection but were highly utilitarian. The adversity experienced in training—emphasized by the life-and-death situation of jumping out of airplanes—built group ties *before* the experience of combat, whereas most regular infantrymen built these ties *during* combat. As much as jumping played a key role, unit pre-combat training regimens proved critical to building esprit de corps. Jumping and training together meant that, despite high casualty rates, paratroopers experienced lower neuropsychiatric casualty rates during the war—about half when compared to regular infantry divisions.[47]

The division, with its distinctive patch or unit insignia, is often cited as the most important echelon with which soldiers identified during World War II. For paratroopers, airborne wings and distinctive uniforms combined with divisional patches to help cement group identity. While every unit in the army experienced unit esprit and cohesion in one way, shape, or form, this comradeship was a degree stronger in airborne units and resulted from a training program designed to create an elite group that expects to fight isolated in enemy-held territory and launch coordinated assaults on objectives within hours of landing.[48]

"There was a powerful feeling of comradeship," recalled John Magill. "In some cases, when it was in the States, it would get a little bit out of hand as our division commanders would sometimes have to clamp down in terms of behavior in town and that stuff. But it was a very, very strong feeling of esprit de corps and being among the best." Chaplain Francis L. Sampson, a recipient of the Distinguished Service Cross for his actions in Normandy, noted the bonding that airborne school fostered. "We sensed, too, that our mutual experience really made us brothers in the airborne family," he wrote in his memoir. Sampson made a significant point that this experience was more extensive than merely the platoon, company, or even regiment—paratroopers felt a bond with the entire airborne community. Jack Nix of the 505th PIR, however, noticed that this inseparable bond between airborne soldiers created a degree of vanity, which he found to be a useful attribute that made these men unwilling to succumb to defeat even when outnumbered or outgunned in their forthcoming battles.[49]

During World War II, the US Army created a new, elite subculture within the service and, through training, imbued it with distinctive values, beliefs, and norms that maintained and enhanced the light infantry mindset upon which airborne culture was erected. Throughout their entire course of instruction, paratroopers endured planned psychological conditioning that convinced them they were "pretty much a super-soldier, mentally and physically." Elitism, flexibility, innovation, decentralization, individuality, and robust leadership created a cohesive force ready for any mission. Some problems arose, including hoarding the most capable leaders in airborne units rather than spreading them around the force. Another problem involved paratroopers treating non-airborne soldiers as inferior, which could lead to fights, especially when fueled by alcohol. However, the army would assume the risk associated with elite units for a return on their investment in the form of increased combat performance. Of course, that is what leaders hoped. The doctrine would still need to be tested in combat.[50]

The values, beliefs, and norms that began percolating in the Parachute Test Platoon of 1940 permeated the entire airborne community. Every newly formed battalion, regiment, or division included trainers drawn from previously created units. Nearly every parachutist passed through the Parachute School at Fort Benning, while those who did not took the same training at other locations. Airborne training was a toughening ordeal that established the paratrooper's courage in the face of danger before entering combat. It also created a new ideal combat leader where "even generals . . . had to be flat-stomached and young." This ideal had lasting consequences for the postwar army.[51]

The first American parachute unit to face the enemy entered combat on November 9, 1942, spearheading Operation Torch, the Allied invasion of Vichy French North Africa. The first regimental-size operation did not occur until July 1943, when the 505th Parachute Regimental Combat Team jumped into Sicily as the vanguard of Operation Husky. American paratroopers performed sixteen parachute operations of varying success between November 1942 and June 1945 as the war continued. The paratroopers acquitted themselves well, and their abilities as regular infantrymen often overshadowed their proficiency as parachutists in both the European and Pacific theaters. Their exceptional performance and the shortage of regular infantry meant that commanders continued placing airborne forces in combat longer than intended, especially during and after the summer of 1944. Their reputation then grew to almost mythical proportions during the final year of World War II.

Meanwhile, at the training bases of the American South, an intense mystique emerged surrounding US paratroopers. Fort Benning served as both incubator and crucible for forward-thinking leaders encouraged by general officers to test new ideas and develop distinct fighting behaviors. Thanks to this effort, paratroopers captured the American public's imagination. From their first prominent feature in *Life* magazine, paratroopers became the US Army's poster children. It was glamorous to "hit the silk." A recruiting poster for the post-Vietnam Special Forces—in part cultural descendants of World War II–era paratroopers—proclaimed that "men join us not because we are different, but because they are." The same was true for the parachute units of World War II, and their test in combat would reinforce a burgeoning airborne culture.[52]

CHAPTER 2

World War II and the Foundation of the Airborne Mafia

> Tough, trained, properly equipped troops, imbued
> with courage and confidence in themselves,
> given sensible and courageous leadership can
> accomplish any of the airborne missions.
>
> —James M. Gavin, *Airborne Warfare*, 1947

World War II served as the essential forma-
tive experience for the airborne mafia. It was in the crucible of combat
that Matthew B. Ridgway, Maxwell D. Taylor, James M. Gavin, and their
subordinates learned how to mold their units into effective fighting
forces, overcome shortcomings, and succeed in the ultimate trial of
war. Through combat experience, airborne units honed a shared set
of assumptions and beliefs about how to solve internal and external
problems, which is the essence of organizational culture. In the caul-
dron of war, the assumptions, values, beliefs, and norms shared among
airborne officers developed into a coherent, collective culture. These
shared experiences likewise validated those precepts already taught
during the training phase. The trial of war—mainly in Europe—helped
frame a unique view of warfighting shared by airborne leaders.

American airborne units performed sixteen parachute "insertion"
operations of varying size and success during the war. From the landings in North Africa on November 8, 1942, to Task Force Gypsy's operation in the Cagayan Valley of Luzon, in the Philippines, on June 23,
1945, parachutists and glider-borne troops developed a unique way
of fighting. Though the Pacific offered some lessons, passed through
reports to Airborne Command at Camp Mackall, North Carolina, it
was in Europe where most of the airborne mafia fought and where the

cultural tenets of decentralization, flexibility, an expeditionary mindset, and adaptability were reinforced through combat operations. Specifically, decentralization and individuality were essential traits in early operations in Sicily and Normandy as airborne units overcame wide dispersion on the drop zone. Stressing flexibility from the junior to senior level helped prepare airborne units to serve well in a rapid-response role. The reinforcement of the beleaguered Fifth Army at Salerno and the response to the German offensive into the Ardennes offered salient examples of the expeditionary mindset. Meanwhile, a capacity for adaptation helped leaders overcome frustration over a lack of mobility and heavy fire support while sowing the seeds for postwar organizational change.[1]

Decentralization

As the paratroopers of Col. James Gavin's 505th Parachute Infantry Regiment boarded their C-47 Skytrain aircraft on the night of July 9, 1943, war was no longer an abstract affair. They were spearheading the Allied effort to invade the island of Sicily, the vanguard for Operation Husky.

Eight months earlier, a smaller airborne operation in North Africa as part of Operation Torch had offered lessons on the likelihood of success. Charged with seizing two Vichy French–controlled airfields at Tafaroui and La Senia south of Oran, Algeria, the airborne mission—dubbed Operation Villain—was ineffective. The fifteen-hundred-mile flight from England through intermittent stormy weather had scattered the aircraft, and only six out of the thirty-nine aircraft carrying Lt. Col. Edson D. Raff's 509th Parachute Infantry Battalion dropped their parachutists on the objective. Most landed thirty-five miles away from their drop zone. By the time Raff's men arrived at Tafaroui, both airfields were in friendly hands, secured by 1st Armored Division tanks. The 509th PIB executed two more parachute assaults in North Africa. On November 15, the battalion seized the Algerian airfield of Youks-les-Bains near the Tunisian border before advancing Germans could use it. Raff's battalion was welcomed with open arms by the French 3rd Zouaves, who presented their distinct badge as a token of partnership. On Christmas Eve 1942, a thirty-man commando-style raid on the bridge at El Djiem failed to locate their objective, and only eight men made it back alive.[2]

After seizing Youks, Raff proceeded to Tebessa with an ad hoc force that included 509th paratroopers, Zouaves, a battalion from the US

26th Infantry Regiment, a company of Algerian *Tirailleurs*, a British an-timine engineer detachment, and a company of tank destroyers. This motley crew next moved to Gafsa and destroyed seventy thousand gallons of fuel before the Wehrmacht could seize it. Known colloqui-ally as "Raff's Army" or "Raff's Tunisian Task Force," the paratroop leader's multinational combined-arms force demonstrated the ability of airborne leaders to operate in combined operations far from sup-port with bold, aggressive action. Maj. William Yarborough, the Fifth Army airborne adviser accompanying Raff, described the officer as a "human dynamo" and wrote that "operating without support or with-out certain re-supply, bothered him none at all." After the raid at Gafsa, Raff's force attacked Faid Pass, a key mountain crossing in southern Tunisia. Raff coordinated a combined-arms air and ground assault on the German-held pass. By the evening of their fourth day, sustained French artillery fire and timely reinforcements finally forced a German surrender. While Raff's actions leading an ad hoc force demonstrate the mindset that paratroopers brought with them into combat, the wide dispersal of aircraft and paratroopers on the initial jump was a harbinger of things to come.[3]

Operation Husky called for the untested 82nd Airborne Division to parachute into Sicily across successive nights. Because of a shortage of transport aircraft, the initial airborne phase of the operation would include only one regiment, fighting as a reinforced combat team. The Allied plan called for the US Seventh Army to land the 3rd, 1st, and 45th Infantry Divisions by sea along the southern coast of Sicily. At the same time, the 505th Regimental Combat Team jumped near Gela to secure high ground in front of the 1st Division's beaches and disrupt enemy reserves. Securing the high ground and road network between the beaches and enemy reinforcements meant capturing "Ob-jective Y," a critical road juncture where the road from Niscemi met the road that ran between Gela and Vittoria. Gavin's 505th was also to assist in securing the landing field at Ponte Olivo for the arrival of Col. Rueben Tucker's 504th PIR and the buildup of Allied combat power, before becoming the II Corps' reserve. The 52nd Troop Car-rier Wing was to fly the 3,405 paratroopers on 266 C-47s in strict nine-ship V formations to ensure accurate delivery of troops closely enough to assemble into fighting units before the moon went down. The second lift, consisting of the 504th PIR, was prepared to reinforce on order. Maj. Gen. Ridgway and his division staff were to wade ashore on D-Day.[4]

Ridgway selected Col. Gavin's 505th because "[he] had done a prodigious job preparing for that attack, and we were ready right down to the last round of ammunition." The 505th Regimental Combat Team consisted of its three battalions plus the 3rd Battalion, 504th Parachute Infantry; the 456th Parachute Field Artillery Battalion; Company B of the 307th Airborne Engineers; and requisite support assets. Of the twelve infantry companies that dropped, only one landed near its objective—Company I of the 505th PIR. During the drop, a lack of training in nighttime formation flying and thirty-five-knot winds resulted in 88 percent of the paratroopers jumping off-target, landing across a sixty-five-mile stretch from Noto to Licata. Most of Gavin's forces landed between the 45th Infantry Division and the German Fallschirm-Panzer Division 1 *Hermann Göring*, which was headquartered at Caltagirone. The 505th PIR spent most of their first day attempting to assemble while attacking and disrupting German and Italian forces wherever they found them. Paratroopers operating in small units destroyed pillboxes, strongpoints, and roadblocks and took hundreds of prisoners.[5]

Initially slated to drop on the night of D-Day, July 10, Col. Tucker's 504 PIR was delayed one night owing to the "obscure situation" of Gavin's forces on the island. When Tucker's men arrived on the night of July 11, the transports carrying them flew over Allied naval forces mere minutes after a Luftwaffe bombing strike. American antiaircraft gunners opened fire, likely believing the Germans had sent another wave of bombers. Twenty-three of the 144 C-47s were destroyed, and thirty-seven others damaged. Most of the troopers jumped, but assistant division commander Brig. Gen. Charles Keerans did not make it out of his plane, becoming the first airborne general officer killed in World War II, alongside eighty-one paratroopers from the 504th.[6]

The paratroopers expected scattered drops, understood the overall mission, and were eager to find and destroy the enemy upon arrival. By noon on D-Day, 2nd Battalion, 505th PIR commander Mark Alexander had assembled a force of 536 men, including 21 artillerymen and one 75mm howitzer. His battalion was mostly dropped together, albeit far from its planned drop zone. He did not, however, expect enemy armor. "Through my field glasses I could see German tanks up the road to the north," Alexander later wrote. "We had been told there weren't any German tanks in Sicily." While Allied intelligence had pinpointed the location of the *Hermann Göring* Division ahead of the operation, to protect the secret code-breaking capability known as ULTRA this information was withheld from the lightly armed paratroopers. Even Gavin had no idea of the strength of the enemy they might face. The men of the 82nd reacted and improvised to defend against an unexpected armored force, proving the efficacy of their training and flexibility.[7]

Lt. Col. Charles Billingslea, attached as an observer, noted in his after action report to the army chief of staff, Gen. George C. Marshall, that "wherever they were dropped, even thirty miles from the D.Z.s, those parachutists went ahead and accomplished valuable missions." And they were doing so for four days—the 82nd was not fully assembled until July 14. The experience in Sicily reinforced to the division's leaders that airborne operations included dispersal and that their efforts to encourage decentralized command and control paid off. The presence of so many pockets of parachutists roaming the Sicilian countryside sowed mass confusion among German and Italian forces. Enemy commanders surmised that multiple divisions were landing in their midst rather than a single reinforced regiment.[8]

Eisenhower's primary airborne adviser, the British general F. A. M. "Boy" Browning, wrote in his July 24 report that the parachutists had caused "widespread alarm, rumors and confusion among the enemy troops," while gauging that the American parachutists speeded up "the landing and advance inland by about 48-hours." Browning had a vested interest in promoting airborne operations yet was often viewed skeptically by Ridgway and Gavin as a rival more interested in supporting British operations than Allied ones. In the same report, Lt. Gen. George Patton remarked that the "swift and successful landings followed by a rapid advance inland would not have been achieved at such a light cost or at such a speed without the action of his airborne division." Little groups of paratroopers strewn about the island no doubt influenced the course of battle, although by how much remains in dispute. In any case, their ability to accomplish their objectives while assisting others, and without assembling large forces, proved remarkable.[9]

To Eisenhower, the supreme Allied commander, the drop seemed a disaster. Wide dispersal, compounded with the offshore fratricide, ensured that Ridgway did not have a cohesive force for four days—hardly what planners had in mind. Eisenhower later remarked that although casualties from the airborne portion of Husky were fewer than feared, the problems in Sicily still caused him to doubt the validity of the airborne division as a concept. Eisenhower recommended that airborne units be employed as no larger than regimental combat teams and not as divisions. Widespread dispersal was uncomfortable for commanders of any unit, even if it was something for which airborne units prepared.

For regimental and division commanders, the promise lay in the potential for increased effectiveness through better training with troop carrier squadrons—that is, the delivery aircraft. If they could land more units closer together, then the possibilities were endless. Regardless of the concerns of senior commanders, the potential of larger airborne units delivered closer together remained promising enough to fuel further developments throughout the war.[10]

The ultimate lesson from Sicily for airborne officers was that training emphasizing individual and small-unit decentralization indeed created effective groups that could accomplish their mission with minimal guidance. The 82nd learned what could and could not be done by parachute troops; rather than spend hours trying to locate command echelons, American parachutists fought the enemy wherever they landed. As Gavin related in his memoir, the fledging paratroopers "learned to move on our objectives immediately on landing; we observed that the first minutes . . . when the paratroopers have the initiative are important to both their survival and the capturing of their objective." Edwin Sayre, commander of Company A of the 505th Parachute Infantry, assessed that the operation was a success because "the major portion of the mission was accomplished" despite the scattered drop. To these paratroopers, Operation Husky proved the efficacy of well-trained small units to sow confusion behind enemy lines, even while dispersed and operating independently.[11]

Concerns raised by Husky, however, spurred development, especially relating to troop transports' navigation, training, and employment. After more testing and training, the War Department, on October 9, 1943, published Training Circular No. 113 (TC 113), *Employment and Training of Airborne and Troop Carrier Forces*. The purpose of the circular was "to provide, in a single reference, information based upon experience gained in recent combat operations concerning the employment of airborne and troop carrier forces." TC 113 highlighted, among other things, the need for "realistic and thorough joint training" and that air routes must be carefully selected and coordinated to avoid friendly naval convoys. The "Knollwood maneuvers" exercise in the winter of 1943 pitted the 11th and 13th Airborne Divisions against each other across four North Carolina counties. The follow-on "Swing Board," led by the 11th Airborne Division commander Joseph M. Swing, studied Operation Husky and the results of the Knollwood exercise and determined that poor operational planning, poor troop carrier training, and piecemeal employment of parachute troops were the primary reasons for the issues in Sicily. The training exercise and Swing's report convinced Army Ground Forces commander Gen. Lesley J. McNair of the efficacy of the airborne divisional concept.[12]

Just two months after the Allied landing in Sicily, the airborne experiment had its next combat test on the Italian mainland. When the Salerno beachhead just south of Naples was in danger of failing in the face of a concentrated German counterattack, the US Fifth Army commander, Lt. Gen. Mark Clark, sent a personal note to Ridgway asking for support (an event explored later in detail). In that message, he also requested that a battalion drop twenty miles inland at the mountain town of Avellino to disrupt German communications and reinforcements. Avellino was a crossroads that, if seized, could significantly affect the Germans' ability to move armor to push the Allies into the sea. Clark requested that this drop, dubbed Operation Giant III, receive priority. Ridgway initially opposed the mission because of the distance the battalion would land from friendly forces, but he reluctantly agreed when convinced that the mission was vital to protecting the British sector of the beachhead. On the night of September 14, the 509th PIB, now commanded by Lt. Col. Doyle Yardley, parachuted into the area from forty C-47s. Navigational errors, combined with mountainous terrain, forced the transports to drop from an altitude of two thousand feet rather than the customary five to eight hundred feet, which badly dispersed the men of the Geronimo battalion over a hundred square miles. Those few who did make it to Avellino could not accomplish their mission. Yardley was wounded, captured, and spent the rest of the war in a prisoner-of-war camp.[13]

Still, 510 of the 640 troopers returned to friendly lines over two months. Working in groups of five to twenty men, the paratroopers fell back on their training, improvised, attempted to avoid detection, and mounted small raids on supply trains, truck convoys, and outposts. Like in Sicily, the overall mission was unmet, but the paratroopers' presence caused confusion and chaos for the Germans. The operation reinforced some principles of airborne operations, primarily that even though little groups of paratroopers can have an outsize effect on the enemy, dropping a highly trained force outside reasonable support distance could yield disastrous results. A decentralized command structure is useful only for so long before units need support, especially massed firepower from ground units.[14]

The Allied airborne assault into Normandy on June 6, 1944, represented another evolution in the decentralized nature of airborne operations. It was to be the largest vertical envelopment yet attempted, with three airborne divisions—two American and one British—dropping simultaneously. In total, 13,056 paratroopers aboard 1,086 airplanes jumped on the night of June 5–6, 1944. The American portion of the drop occurred in two assault lifts in close succession: Mission Albany (101st Airborne Division) and Mission Boston (82nd Airborne Division). The 82nd and the 101st fought in ad hoc units as enemy fire and clouds scattered their transports, preventing units from consolidating, as expected. Once again, dispersal was mitigated by emphasizing decentralization. In Sicily, paratroopers had been "instructed to attack the enemy, regardless of the size of the opposition, wherever they met him"; in Normandy, they were instructed to move at night and avoid large enemy groups. Individuals assembled with men from different companies, battalions, regiments, and even divisions; deliberate assembly was all but impossible. As the 505th Parachute Infantry was the only regiment with combat experience that jumped in Normandy, paratroopers from that regiment often took charge of small ad hoc groups.[15]

The missions for each division were basic airborne tasks designed to facilitate the seaborne landings. The 82nd was directed to seize and hold river crossings at La Fière and Chef-du-Pont, clear enemy forces along the Merderet River, capture Sainte-Mère-Église, destroy crossings over the Douve River at Beuzeville-la-Bastille and Étienville, and prevent the movement of any enemy reserves into the beachhead area. The 101st Airborne Division mission called for their 6,600 men to seize the exits of four causeways that led inland, to assist the 4th Infantry Division's landing at Utah Beach and likewise secure the VII Corps' southern flank along the Douve River. The causeways were four elevated roads that constituted the only paths off the beach after the Germans prepared their defenses on the Cotentin peninsula.[16]

The 101st Airborne Division's drop pattern was enormous—fifteen-by-twenty-five miles—yet more than 70 percent of the division landed within an eight-mile square. As the seaborne forces were landing, the division only had 1,100 men organized but rapidly took its objectives. By midnight at the close of D-Day, this number had increased to 2,500. Most fought in small groups under whatever leaders were present while moving on various division objectives. One platoon leader in the 506th PIR remembered that he never had more than half his platoon at any given time. Over 60 percent of the division's equipment was unaccounted for, and division commander Maj. Gen. Maxwell Taylor worried

FIGURE 2. Sgt. Joseph F. Gorenc from Sheboygan, Wisconsin, of the 506th PIR, 101st Airborne Division, climbs aboard the lead transport aircraft C-47 Dakota 8Y-S "Stoy Hora" of the 440th Troop Carrier Group at RAF Exeter Airfield, Devon, England, on the night of June 5–6, 1944. The paratroopers would drop behind Utah Beach on France's Cotentin peninsula, near Cherbourg. US Army photo, RG 111, NARA.

that "the division could not have maintained itself much over 24 hours without support." Included in that figure of missing equipment were eleven of twelve airdropped howitzers from the 377th Parachute Field Artillery Battalion. Despite the scattered landings, the division took its objectives, including Varreville and Pouppeville, and held the critical causeways leading from Utah Beach to allow the 4th Infantry Division to move inland. By shortly after noon on D-Day, elements of the 3rd Battalion, 501st PIR, established the first link-up between air and seaborne forces.[17]

The 82nd Airborne Division had similar problems with its drop. Scattered over a four-square-mile area, the division could account for only 30 percent of its personnel by the end of the day on June 6. By June 8, the division had only 2,100 effectives. However, the division accomplished its mission by June 10 (D+4). By the end of the campaign, the 82nd Airborne Division had suffered 5,363 casualties: 778 killed, 3,373 wounded, and 1,212 missing—all but nine missing from the initial landings. This was from 11,657 men who took part, or a 46 percent casualty rate for the entire division and 54 percent among infantry units. Of the three regiments that landed, the experienced 505th PIR was given the most challenging task and assembled quickly. By 3 p.m. on D-Day, the 505th had more than 1,100 men assembled of the more than 2,000 who jumped, and by the evening had linked up with Company A of the 8th Infantry Regiment from the 4th Infantry Division landing at Utah Beach. The 507th PIR, on the other hand, landed mostly west of the Merderet, and not one single stick (planeload of paratroopers) landed on the drop zone. They were unable to operate as a regiment until June 9. The 508th PIR experienced similar problems, and for the first two days, only one of its three battalions established contact with regimental headquarters.[18]

The scattered drop in Normandy reinforced airborne leaders' belief in the ability of well-trained and well-led decentralized small groups to accomplish unit objectives. Often, the groups securing planned objectives were roughly one-fifth the size of the element planned. After four days, the 82nd finally held both sides of La Fière bridge—vital for linking up with isolated pockets of paratroopers west of the Merderet River. The 101st had succeeded in securing not only the causeways leading to Utah Beach but also the lock system at La Barquette, preventing enemy reinforcements from threatening the seaborne forces on Utah Beach. While ultimately successful, the airborne forces had not met their objectives on time, as all missions were supposed to be accomplished by dawn on D-Day. Still, Eisenhower seemed pleased and wrote to the Combined Chiefs of Staff on D+4 that "the landings on the Cotentin Peninsula apparently went about as well as could be expected with the 101st Airborne Division carrying out its missions in good style. Information on the 82nd Airborne is meager, but General Bradley informed me that the VII Corps had made contact with it." While subject to the expected problems with dispersion, airborne units relied on their training, and their relative success despite the scattered drops helped garner fame for themselves and their commanders.[19]

Casualties among the paratroopers were high, especially among the infantry regiments, but nowhere near the estimate of Trafford Leigh-Mallory, the British commander in chief of the Allied Expeditionary Air Force, who had predicted a staggering 75 percent casualty rate. In the 82nd, total casualties from thirty-three days in combat were 46 percent killed, wounded, or missing—5,245 men. The three parachute infantry regiments suffered 55 percent dead or wounded, and the 325th Glider Infantry suffered a staggering 58 percent casualty rate. Tactical leaders suffered heavily as well; of the twelve commanders who led 82nd Airborne battalions on D-Day, only one emerged unscathed—two killed and four wounded, the rest injured or relieved. The 101st suffered 4,670 total casualties, or 39 percent of the division, including its assistant division commander, Brig. Gen. Don Pratt, in the glider assault on D-Day. The "Screaming Eagles" of the 101st also lost two regimental and five battalion commanders during the fighting in Normandy.[20]

The disorganized nature of the drop was not, according to the postwar Weapon Systems Evaluation Group, "an unmitigated evil, and in fact, contributed to a degree to the general success of the invasion." With so many small teams of paratroopers roaming the Normandy countryside, German defenders had no significant, worthwhile targets to attack—many German prisoners reported hopelessness when defending against such an airborne attack, as vertical envelopment presents multiple dilemmas to enemy forces. The chief of staff of the German 7th Army noted the effect of airborne troops on preventing German troops from reaching the beachhead; the drop had in fact succeeded in screening enemy forces in front of the beaches. Echoing official and prisoner reports in his report for the Advanced Infantry Officer course in 1949, 506th staff officer Salve Matheson attributed mission success to unit dispersion and pondered whether the drop would have been as successful if executed as planned.[21]

The drop into Normandy also reinforced many lessons regarding assembly and the principle of mass. Rather than paratroopers fighting alone or in small groups, complete planeloads of men should assemble their equipment as promptly as possible to ensure a maximum mass of manpower and firepower. Yet all paratroopers were to expect dispersion and realize that no unit would ever assemble all its men, so leaders must be prepared to move on objectives with less than the expected number of personnel. This idea was juxtaposed with the need for individuals and small units to take prompt, decisive action as soon as possible to maintain the initiative. After Sicily, though, leaders learned to exercise caution when assembling and, rather than attacking the enemy immediately, to avoid large enemy groups until enough men were assembled.[22]

The efficacy of dispersed airborne operations—small units operating behind enemy lines with minimal support—was proven in Sicily and Normandy, thanks to an emphasis on decentralized leadership to mitigate the effects of dispersion. Factors outside the control of the airborne troops, such as fog, air defenses, and nighttime navigation errors, contributed to widespread dispersion during these operations. Owing to poor navigational capability and technology, night drops hampered airborne units' ability to assemble and link up with ground units. Accordingly, American planners abandoned this tactic in September 1944 in favor of daylight operations. The airborne portion of the invasion of Southern France likewise featured decentralized success amid a much more coherent drop. Sixty percent of First Airborne Task Force paratroopers landed on their drop zones. One group of aircraft jumped early, however, and the paratroopers from those twenty planes captured critical antiaircraft and coastal batteries in the 3rd Infantry Division's zone at St. Tropez. Operations Market Garden and Varsity, later in the war in the Netherlands and Germany, and smaller operations in the Pacific achieved more accurate drops, thanks to daylight conditions, Allied control of airspace, and concomitant suppression of Axis defenses.[23] Yet the earlier operations in Sicily and Normandy validated that well-trained small units can sow confusion behind enemy lines. The lessons learned were instrumental in preparing the new airborne units from the United States. But so were lessons in Italy and later at the Battle of the Bulge about the employment of airborne troops as a mobile reserve.

An Expeditionary Mindset

It was two months after the invasion of Sicily and a little less than nine months before D-Day in Normandy when Allied forces stormed the beaches at Salerno, Italy. Beginning at about 3:30 a.m. on September 9, 1943, two corps—the American VI and British X—landed five divisions amid light German defenses, but German reinforcements quickly converged on the beachhead. The Allied forces faced stiffer resistance than in North Africa and Sicily, and Fifth Army commander Lt. Gen. Mark Clark's beachhead was in peril by the 12th. He had already committed his floating reserve, the 45th Infantry Division, and had run out of options. The German 10th Army under Field Marshal Kesselring kept intense pressure on the beleaguered Allied beachhead, and "it looked as if the German attack might drive clean through to the beach," as Gavin later recalled. The 82nd Airborne Division was sitting in reserve in Sicily at the time; it was available only thanks to the cancelation of an ill-conceived plan to seize four airfields around Rome.[24]

Fearing he would lose the beachhead, Clark activated his emergency reserve force by sending a personal dispatch to Ridgway on September 13, asking for immediate help. "Since the German counterattacks were becoming steadily stronger," Clark remembered, "I was only too happy to have the paratroops available." Gavin received the pilot who brought the note, recalling, "The afternoon about 1330, a tired, begrimed pilot landed at Licata Field in Sicily in a fighter from the Salerno beachhead. He had an urgent message for the division commander and refused to give it to anyone else." Ridgway had just left Licata for Termini, but the division chief of staff recalled the plane to enable the courier to deliver the message. Clark's note asked for one regimental combat team to jump onto the beach that night and assemble at the Paestum ruins south of the Sele River.[25]

Ridgway sent back a terse reply, "Can do," and within two hours, the 504th Parachute Infantry Regiment was assembling near their aircraft. Using C-47 tails as map boards, the leaders briefed the mission planeside. The division then shuffled men to different departure airfields, finalized plans, and coordinated with friendly units before loading paratroopers. Clark also relayed plans for marking the drop zone, which his airborne adviser Maj. William P. Yarborough put together. The marking plan was ingenious and included lighting gasoline-soaked cans of sand in a large T on the beach. This marking system was possible because of the location of the drop zone—behind friendly lines.[26]

Within eight hours, forty-one C-47 Skytrains of the 52nd Troop Carrier Squadron containing the paratroopers of Col. Rueben Tucker's 504th PIR lumbered down runways in Sicily. Shortly after midnight, the regimental combat team landed, assembled, and was "looking for a fight" by daylight. Most paratroopers landed within two hundred yards of the drop zone. Tucker—whom Clark described as "a real fighting soldier"—reported to the Fifth Army headquarters at 3 a.m. on the 14th. "As soon as assembled, you are to be placed in the front lines," Clark told Tucker. "Sir," Tucker replied, "we are assembled and ready now." By daylight, fifteen hours after Clark's request, 1,300 paratroopers were on the ground under Tucker's command. The most effective use of paratroopers to this point in the war was in this role as a highly mobile reserve.[27]

On September 14, while the 509th Parachute Infantry Battalion dropped in the hills around Avellino, Gavin's 505th PIR entered the fray at Salerno, giving Clark two entire airborne regimental combat teams on the beachhead. The 82nd Airborne Division then prepared to send in gliders from Sicily containing the 32nd Glider Infantry Regiment on the night of the fifteenth but found no suitable landing areas, and that regiment landed by sea instead. The effect of the airborne soldiers on the battle's outcome was significant and provided a needed boost in morale and manpower at a critical point. By September 17, the Allied forces closed the gap along the Sele River and attacked inland, led by the 82nd Airborne, using heavy artillery from the regular infantry divisions. The 504th PIR, led by the aggressive Tucker, who Gavin surmised might have been the best regimental combat commander in the entire war, seized the village of Altavilla and its commanding high ground. After a staunch forty-eight-hour defense, the Germans withdrew, and the beachhead was never threatened again.[28]

As in previous operations, airborne division organic fire-support assets proved insufficient; 155mm heavy artillery temporarily brought over from regular infantry divisions was crucial to success in Southern Italy. Nevertheless, the quick insertion of airborne troops significantly impacted the Salerno beachhead. Airborne officers, of course, felt they were decisive in the Allied victory. More important was the realization that, after being committed two hundred miles away within twelve hours of notification, airborne troops had demonstrated an essential capability as a mobile striking force "that no high commander could overlook in the future." Throughout the Mediterranean campaign, airborne units had shown their suitability for seizing key terrain and disrupting enemy reserves. Because of an emphasis on flexibility in training at all levels, the 82nd's action at Salerno demonstrated that paratroopers could also rapidly reinforce a threatened operation, thus strengthening the idea that parachute-delivered forces could be valuable in the future.[29]

When the Wehrmacht launched its offensive into the Ardennes forest fifteen months later, in December 1944, the XVIII Airborne Corps, commanded by Lt. Gen. Matthew Ridgway—promoted from division to corps command—sat in reserve for the Supreme Headquarters, Allied Expeditionary Forces. Eisenhower tasked this force with countering the advancing German onslaught in what became known as the Battle of the Bulge. The airborne reserve, under the circumstances, had neither the time, transports, nor favorable weather to make a parachute assault to reinforce the beleaguered Allied line. Instead, they would become "truckborne" infantry, making a tailgate jump from the army's standard 2.5-ton trucks. (Later, the image of valiant paratroopers unprepared for the weather, scrounging ammunition on their way into battle, would be etched into the collective national memory thanks to the 1949 film *Battleground* and the 2001 miniseries *Band of Brothers*.) The Germans were pushing toward the Belgian town of Bastogne with one reinforced division, while significantly larger German forces bypassed the town and continued toward the Meuse River. Plugging the gap and responding to the German attack toward Bastogne was crucial.[30]

In northeast France, the 82nd and 101st Airborne Divisions were in varying states of readiness. Neither division was prepared to go into combat on such short notice. Just one month earlier, Ridgway had informed his two subordinate division commanders—Gavin and Taylor—that their units would have until February 15, 1945, to reconstitute, with the caveat that "there is no guarantee that [the time] may not be shortened." Many units had turned in their weapons for much-needed maintenance. The paratroopers of both divisions moved out in disarray, with some in summer-weight uniforms they had worn in Holland, some in dress uniform, and some in winter coats recently received. The lack of winter clothing was not new; Taylor had raised this concern to Ridgway as early as November 7. And commanders were worried about replacements entering combat so quickly after minimal training and integration. Bill Guarnere of the 506th PIR remarked later, "It made you nervous to have so many replacements going in. A lot of the original company was gone, and you have no idea what these new kids are going to do." Rumors abounded as well. Pvt. Vincent Speranza, a replacement in the 501st PIR, remembered that "all kinds of wild theories flew about. Eisenhower had been captured, and we had to rescue him. We were being moved up to be the first ones in Berlin. The Germans had surrendered, and we were to escort all the prisoners back to France."[31]

When the XVIII Airborne was alerted, Ridgway and most of his staff were in England learning to work together and preparing Maj. Gen. William Miley's green 17th Airborne Division for combat. Ridgway gathered his staff and got on what turned out to be the last C-47 flying across the Channel before the weather set in and prevented further flights and resupply. As soon as the weather allowed, Miley's division would follow, to buttress Patton's Third Army forces.

Acting as the corps commander in Ridgway's absence, Gavin moved his 82nd first, followed by the 101st. Brig. Gen. Anthony McAuliffe was in command of the latter division, as Taylor was in Washington discussing with General Marshall the best ways to organize and employ the airborne division for the war in Europe. As Gavin planned defensive positions with the First Army's Lt. Gen. Courtney Hodges, he decided to push the 82nd farther, to Werbomont, and have the 101st stop at Bastogne. The Screaming Eagles arrived just eight hours ahead of the Germans. This decision to defend Bastogne was made in haste, owing to geographic circumstances, but had lasting effects on the popular legacies of both units. Even while the two airborne divisions were moving toward the salient, Ridgway had arrived in France and followed the last units out of their bivouac areas. Meanwhile, tanks from the 10th Armored Division raced north to Bastogne to reinforce the 9th Armored Division elements already on site.[32]

On the northern edge of the enemy thrust, the 82nd Airborne Division endured, in Ridgway's words, "less publicized, but equally as severe" fighting as what would be faced by its sister division at Bastogne. In the initial action, the division deployed its constituent regiments to defensive positions at the many crossroads near Werbomont, outside of mutually supporting distance. Until the 30th Infantry Division and elements of the 7th Armored Division joined Ridgway's corps, there was little choice, and still, the division performed well. Gavin's leadership helped keep morale up when the weather seemed to doom all hope of survival; one trooper related a chance encounter with Gavin on patrol in which he told the general he was having a difficult time keeping his feet dry. Gavin "took off his overshoes and gave them to me," the trooper said. Despite having fractured two vertebrae in the parachute assault into Holland, Gavin shared his men's hardships—visiting men in their foxholes throughout the winter—and his men revered him for that.[33]

Gavin was not the only story of toughness in the Bulge. Two soldiers from the 82nd Airborne Division helped create a legend that bolstered the division's pride for decades. Pfc. Thomas Martin and Pfc. Vernon Haught were members of the 325th Glider Infantry near St. Vith, Belgium. On the morning of December 23, 1944, a tank destroyer from another unit approached Martin's foxhole, looking for the front line. Martin boldly told the crew, "Buddy, just pull your vehicle behind me.

I am the 82nd Airborne, and this is as far as the bastards are going!"
This boast became famous, though it was sometimes misattributed to
Haught because of the iconic photo of Haught from the battle. The
two stories merged, and the quote and image were used on a reenlistment poster. Martin and Haught embodied the gritty, cocky attitude
shared by many in airborne units. Their steadfast defense at St. Vith
highlighted the tenacity and bravery of the 82nd Airborne Division in

FIGURE 3.   Pfc. Vernon L. Haught, from Payson, Arizona, of the 325th Glider Regiment, 82nd Airborne Division, after three hours on guard at an outpost, returns for a little rest and heat, near
Ordimont, Belgium. US Army photo.

the face of fierce German attacks. The episode embodied the airborne spirit that swelled division pride.[34]

Upon entering the line, Ridgway and the XVIII Airborne Corps assumed command of the 30th Infantry and 3rd Armored Divisions. During the operation, Ridgway commanded six subordinate divisions at one point, and his corps played an instrumental role in halting the German advance by preventing the enemy from turning the First Army's flank. The entry of his corps into the line provided much-needed optimism for Lt. Gen. Hodges and the First Army staff. Ridgway was an intense, charismatic leader—bold and aggressive—who led from the front like Gavin. He and Gavin both believed they should launch an attack immediately after they learned the German disposition. "It seemed most important to plan an *attack* the moment the German penetration was checked," Ridgway wrote. On December 19, the 82nd mounted an attack alongside tanks from the 7th Armored Division. The bold, aggressive action taken by the 82nd steadied the front and began to return the initiative to the Allies.[35]

In Bastogne, the 101st joined with about a division's worth of other American forces and fell under Maj. Gen. Troy Middleton's VIII Corps. Combat Command B of the 10th Armored Division, the Combat Command Reserve of the 9th Armored Division, the 705th Tank Destroyer Battalion, and two 155mm artillery battalions—the 755th and the African American 969th from VIII Corps—strengthened their defense. Also included were scattered elements of the 106th and 28th Infantry Divisions organized into a makeshift unit called Team SNAFU. The 101st had beaten the Germans to the town, and McAuliffe deployed his infantry regiments to defend the primary approaches and sent men from Lt. Col. Julian Ewell's 501st Parachute Infantry to "develop the situation" by reconnoitering forward. McAuliffe likewise organized the limited number of tanks and tank destroyers into a mobile reserve force, ready to reinforce hard-hit portions of the defense as needed.[36]

After deploying reconnaissance patrols, Ewell's men encountered exhausted German advance elements on December 19. The fresh, motivated Americans in control of the town provided a stark contrast to the tired, ill-supplied Germans who thought their numerical advantages should have been sufficient to sweep aside the American defenders. Instead, the Germans only managed to encircle the defenders of Bastogne on December 20 as the German advance continued to push west on both sides of the town. American paratroopers, trained to operate behind enemy lines without heavy support, did not allow the uncertainty of the situation to faze them. With artillery and armor support this time, the 12th Army Group commander Omar N. Bradley was sure of their ability to defend the critical road junction. "I was confident that the 101st could hold with the aid of those tankers from the 9th and 10th Armored Divisions," Bradley wrote later. When a German messenger came to McAuliffe's command post with a note demanding surrender, McAuliffe sent back simply, "Nuts!" as his reply. This rhetoric came at the urging of his division operations officer, Lt. Col. Harry W. O. Kinnard, who recommended that McAuliffe's first reaction (he reportedly said "Aw nuts" when reading the German note) was sufficient. Nevertheless, the defenders' relief became a priority for Bradley: "They could hold out I thought at least until Patton's Third Army broke through to relieve them. The relief of Bastogne was to be the priority objective in Patton's flanking attack."[37]

The 101st held Bastogne, outnumbered and surrounded, for five days before Patton's Third Army, led by Lt. Col. Creighton Abrams's 37th Tank Battalion of the 4th Armored Division, broke the German siege. Contrary to popular memory, the 101st was never alone. Newspaper reports covered the siege and subsequent breakthroughs in detail, helping tell the story of the 101st Airborne Division's heroics and Patton's breakthrough while often ignoring the efforts of other units during "the Bulge," much to the chagrin of the supreme commander, Gen. Eisenhower. Reports in the *New York Times* continued to reinforce the unlikely heroism of the 101st—including a story that the division did not need rescuing. Even Eisenhower acknowledged the speed and impressiveness with which the two airborne divisions moved. "The momentum of the thrust was further reduced by the arrival of two airborne divisions moved from reserve in the Reims area on 18th December," he wrote to the Combined Chiefs of Staff; "one of these (reinforced by armor), although under constant attack and completely surrounded for five days, held the important road center at Bastogne." These newspaper stories did much of the heavy lifting in developing the legend of the 101st Airborne Division and its stand in the Ardennes.[38]

The 101st's staunch defense and McAuliffe's refusal are critical components of unit, army, and national memory. Missing the first week at Bastogne was one of Maxwell Taylor's two significant disappointments of the war. However, he consoled himself in the knowledge that it gave "McAuliffe a chance to become a world figure." The ability of beleaguered but determined paratroopers to defend against a mechanized onslaught helped portray airborne forces as effective against all types of enemy units while contributing to mythology about their capability. What the broad legend fails to encapsulate, however, is the amount of support given to the 101st by non-airborne elements, especially tanks and heavy artillery. Besides emphasizing the capability of airborne units to move quickly in crises, the fight at Bastogne reinforced the requirement that airborne units be equipped much like standard infantry divisions and the role of tank support. "Somehow we airborne foot soldiers seemed to bring out the best in the armor and they seemed to bring out the best in us," Kinnard remembered. Against an enemy mechanized force, friendly armor and tank destroyers were instrumental to success. "There were a lot of tanks just scattered around. They were doing most of the shooting," according to Lt. Eugene Drance. Despite the combat record of airborne units, their élan, and their stubbornness, the fight in the Ardennes displayed how minimally equipped light infantry depended on armor and artillery.[39]

The actions on the Salerno beachhead and the bolstering of Allied lines in the Ardennes offered crucial lessons about rapid reinforcement, flexibility, and leadership. The airborne troops displayed a penchant for flexibility and innovation throughout these battles, while leaders showcased courage and care for their men. Despite lacking tactical mobility once on the battlefield, airborne units had immense operational reach. Their expeditionary design emphasized lightness and air transportability, making them as suitable for sea or land movement as by air, ideal for a quick-response role. Adapting to changing situations is critical for success in any unit. This adaptability was a deliberately cultivated skill during stateside training for airborne units. By stressing streamlined units and readiness to move at a moment's notice, leaders created a dynamic rapid-response force that instilled an expeditionary mindset.[40]

Adaptation

Much like coping with problems of equipment shortages and a lack of aircraft in training, the ability to adapt and improvise in combat proved key to overall mission completion. For airborne forces in Europe, challenges included their paltry ground mobility, lack of firepower, and the unsuitability of glider transports. Airborne unit organization, equipment, and training stressed lightness and air transportability—organizational choices that reflected a flexible, self-reliant mindset that stressed decentralized leadership, as airborne troops needed to fight rapid wars of movement. However, these decisions made the airborne division, in some circumstances, highly immobile and very vulnerable. Follow-on forces were presumed to bring heavier weapons and transportation, yet the light airborne divisions often fought longer than the doctrinal seventy-two hours after they landed. Airborne divisions were counted on to fight like infantry divisions yet contained far fewer transportation and fire-support assets.[41]

Higher headquarters organized and attached elements to airborne divisions throughout the war, allowing them similar capabilities as regular infantry divisions. Airborne small-unit tactics were always meant to be the same as those of the regular infantry—airborne infantry units are, after all, staffed by infantrymen—but problems arose with sustained combat. The postwar General Board found that the insufficient personnel, transportation, and firepower "placed an extremely heavy burden on the airborne division" and contributed to the inability of these divisions to sustain themselves for more than a few days without support. The structure and assigned firepower for the divisions were inadequate for the time they spent committed to combat. The airborne division was *too* light. It contained too few howitzers, too few trucks, and no tanks or tank destroyers.[42]

The 82nd did not gain organic parachute-delivered artillery until Ridgway created a parachute field artillery battalion at the army's behest a month before departing the United States. Before the advent of parachute-delivered artillery units, howitzers were meant to arrive by glider. Most airborne division artillery consisted of 75mm "pack" howitzers, so named because their original design was for mountain operations—a cannon that could be disassembled for transport on multiple pack animals. Thanks to the ability to disassemble the guns, they were ideally suited for a piecemeal parachute drop. Airborne divisions had no assigned 155mm artillery, whereas regular infantry divisions had twelve of the larger howitzers. Airborne divisions also had only twelve 105mm artillery pieces, while regular infantry divisions had fifty-four. According to Maj. Edwin Sayre, "Airborne troops are not capable of sustained action unless quickly reinforced by artillery comparable to that of the enemy." Ridgway remembered that in Normandy it was not until his division contacted conventional units and gained the use of their artillery that they "could get on apace with our basic mission." Further, airborne divisions had twice the number of M1919 .30-caliber machine guns rather than heavier M2 .50-caliber machine guns, owing to a lack of transportation to carry the 128-pound behemoth known as "Ma Deuce." The lack of heavy machine guns hampered the airborne division's ability to sustain a fight against armored vehicles and led to an increased reliance on attached tank and tank destroyer units.[43]

Fire-support capability was an enormous and distinct advantage for American forces throughout the war and became, over time, the best in the world. The capability of American artillery in a firepower-centric army, coupled with the lack of fire support organic to airborne divisions, made external fire support a constant demand for airborne units. In North Africa, attached French artillery assisted Raff's force in accomplishing its missions. Taylor remembered during operations in Sicily that his "light airborne artillery was reinforced by the attachment of the 155mm howitzer battalion of the 9th Division commanded by a Lieutenant Colonel W. C. Westmoreland, whose surehanded manner of command led to the entry of his name in a little black book I carried to record the names of exceptional young officers for future reference." In his post-Sicily analysis, Maj. Gen. Joseph M. Swing asserted that with increased artillery and transportation, the airborne divisions were suitable for extended action akin to standard infantry divisions. At the same time, Gavin realized there was no need to alter the force structure for units to make them airborne, so long as adequate air transport was available to fly them to the objective. Airborne units could have used air-transportable, heavier artillery and vehicles, but the technology did not support such advances at the time.[44]

Organizational issues arose as well. The entire force structure was heavily streamlined to maximize air transportability, which was key in crafting an expeditionary mindset but provided problems when the airborne fought as regular infantry for extended periods. In addition to a lack of firepower division-wide, each rifle parachute platoon had just two rifle squads. Moreover, the original table of organization and equipment (TO&E) had each airborne division organized with two glider infantry regiments and one parachute infantry regiment. Following the Normandy campaign, in which gliders exhibited significant vulnerabilities, the TO&E was formally changed to one glider and two parachute regiments. In the 11th Airborne Division, where there remained two glider regiments and one parachute regiment, the division commander trained his units to use both means of arrival. Maj. Gen. Swing learned early that glider operations were useless in a jungle environment. He trained all glidermen as qualified parachutists, turning his three infantry regiments into "para-glider" units in a harbinger of changes after the war.[45]

Ridgway corresponded with Marshall consistently in regard to the overall size of airborne divisions throughout the war, lamenting that the original streamlined authorization was insufficient for more than three days in combat. In Normandy and Holland, the 82nd and 101st unofficially operated with 13,000 to 14,000 men across four infantry regiments; in Ridgway's belief, the airborne division organization needed to be officially and permanently increased to the size of a standard infantry division. When Taylor was back in Washington in December 1944, he discussed this issue with Marshall. The army's chief of staff listened and granted an increase in airborne division strength. Airborne officers learned to solve operational problems by altering force structures to match reality.[46]

Glider operations represented another source of frustration and learning for airborne leaders. Speaking well after the war, Taylor lamented that the glider was the worst way to arrive on the battlefield. Early airborne theorists foresaw parachute troops as an arrowhead to clear the way for gliders intended to deliver massed combat power and heavier equipment behind enemy lines. Nevertheless, glider operations never came into their own, despite the robust effort. A shortage of equipment precluded the airborne command's ability to prepare an adequate number of pilots and men; hence there was a two-platoon and two-battalion table of organization in glider infantry regiments. Furthermore, while the original division organization contained two glider regiments, a lack of shipping space on the cargo vessels headed to North Africa necessitated that the 82nd Airborne Division leave the 326th Glider Infantry Regiment (GIR) in the United States. Gavin's 505th PIR joined the division as its replacement.[47]

Significant cultural issues caused problems as well. Glider infantrymen were nonvolunteer personnel, looked down on by the all-volunteer parachutists for their lack of jump wings. They flew in big, slow, lumbering, unpowered aircraft that were easy targets for enemy antiaircraft gunners. And the gliders were fragile. Glider pilots also notoriously shirked ground duties, as the army failed to assign them tasks as part of a unit upon landing. Glider infantry regiments had one fewer platoon per company and one fewer battalion per regiment than standard or parachute infantry regiments. For Normandy, the divisions handled the problem by splitting the two battalions of the 401st GIR between them, giving the 325th and 327th GIRs a third battalion each. However, the ability of glider-delivered field artillery and antiaircraft units to provide larger amounts of firepower was essential to success throughout the war. Antiaircraft units such as the 80th Airborne Anti-Aircraft Battalion of the 82nd Airborne Division were routinely employed in an antitank role, especially after the Allies achieved air superiority.[48]

Normandy was the first action in which glider infantry regiments were committed to combat via glider assault. Gavin credited the aerial resupply via parachute and glider on D+1 as "decisive in enabling the 82nd Airborne Division to hold its airhead until it could make contact with the amphibious landing force." (Like a beachhead, an airhead is the portion of territory under the invading unit's control that can receive more personnel and equipment, in this case by air.) However, the 325th GIR suffered sixteen men killed and seventy-four injured during the landings alone. Fifty-five percent of the 82nd's Horsa gliders were demolished, and 10 percent of personnel riding in them were injured, compared to 6 percent of CG-4A glider riders. The Horsa, developed in the United Kingdom, was twenty feet longer and carried up to twenty-eight troops, compared to the thirteen-troop capacity of the American-built CG-4 Waco. And as was the case for the parachute troopers, inadequate transport-pilot training resulted in increased casualty rates. Teddy H. Sanford, commanding the 1st Battalion, 325th GIR, in Normandy, commented that rather than coming in slow from a high enough altitude to make a proper approach, the pilots "had no opportunity for selection of the field, or to turn to make any approach to it. It was just cut loose and land, which put a great many of our gliders into the trees and resulted in rather high casualties." In Normandy, 6.8 percent of the troops committed by glider necessitated evacuation to England, as opposed to 3.8 percent of those committed by parachute. This discrepancy helped Ridgway, Taylor, and Eisenhower lobby Marshall for increased pay for glider infantrymen, which was equalized with parachute pay in July 1944.[49]

Poor glider landings in the 101st Airborne Division resulted in the loss of their assistant division commander, Brig. Gen. Don F. Pratt, on the morning of D-Day. While the 82nd sent most of the 325th by glider, the 101st's 327th GIR arrived by seaborne landing, owing to a shortage of gliders. Ninety percent of the four thousand troops inserted by gliders arrived safely, but more than 50 percent of the equipment delivered by gliders was destroyed. Members of the 82nd, for example, recovered only eight of sixteen 57mm antitank guns of the 80th Airborne Anti-Aircraft Battalion and few of the jeeps earmarked to tow the weapons around, meaning that often paratroopers were pulling them by hand. Normally assigned to antiaircraft duties, the 80th was outfitted with antitank guns to provide an extra punch for the lightly armed airborne division. Undoubtedly, the ability to deliver equipment via glider was instrumental in giving airborne units a fighting chance in Normandy before they linked up with seaborne forces, armor, and heavier artillery. Accordingly, the idealized image of the glider remained. Glider-delivered 75mm artillery, 57mm antitank guns, and transportation to haul them around the battlefield were critical to the units' successes. Gliders "carried 95 howitzers and anti-tank guns, 290 vehicles, 238 tons of cargo, and 4,021 men into Normandy" across five separate missions on June 6 and 7, 1944. Nevertheless, because of the rate at which gliders crashed and the casualties incurred, gliders remained subordinated to parachute operations. Ninety-seven percent of the gliders used in Normandy were never recovered.[50]

Glider operations improved throughout the war. During Operation Market Garden—the ill-fated Allied attempt to cross the Rhine through Holland in the fall of 1944—gliders were not part of the initial assault on September 17. Market Garden called for an airborne assault on three critical bridge crossings by three Allied airborne divisions (Operation Market). This was followed by an armored thrust along a two-lane road to push into Germany (Operation Garden). The American 101st landed near Eindhoven, the 82nd farther north at Nijmegen, and the British 1st Airborne Division landed even farther north on top of the remnants of a German SS Panzer division at Arnhem. The British division landed too far from its objective, while the ground force moved too slowly to reach the furthest committed airborne elements before their destruction. After the initial parachute assault, 1,890 gliders flew into Holland over the ensuing seven days with reinforcements and resupply. Of that number, 1,530 landed without incident on or near their assigned landing zones, compared with 293 that did not, while 67 were unaccounted for—a marked improvement from Normandy. Even better, the Americans recovered 95.9 percent of glider-delivered resupply, compared to 41.4 percent of that delivered by parachute. Glider-borne resupply also played a vital role in Bastogne, delivering 139,281 pounds of cargo on December 26 and 27. Among the gliders' most precious cargo in the Ardennes were desperately needed surgical teams for the more than four hundred casualties that required care after most of the 101st's medical personnel were captured before the encirclement of the division.[51]

During Operation Varsity on March 24, 1945, the 194th Glider Infantry Regiment's 906 gliders began landing at 10:30 a.m. The overall mission of Operation Varsity was to secure a crossing over the Rhine near Wesel, Germany, as the airborne component of Operation Plunder. The mission was so successful that Gavin called it "the highest state of development attained by troop-carrier and airborne units." Of 572 gliders launched on Varsity's D-Day, only 50 were destroyed, and the rest delivered 3,492 men, 202 jeeps, 94 trailers, and various amounts of munitions and artillery. Only 83 gliders landed outside the landing zone, and the 194th Glider Infantry assembled at 75 percent strength by noon. The regiment accomplished its mission to hold crossings over the Issel River, and the division achieved all its objectives by nightfall. Gliders were also used as medical evacuation for the first time in the European Theater. Mimicking a technique used in India and Burma, Lt. Col. Robert Burquist, chief surgeon for IX Troop Carrier Command, had gliders converted into air ambulances with medical personnel on board. Two of these retrofitted gliders landed beside the First Army's field hospital on the east bank of the Rhine and loaded casualties. Minutes later, the tow planes returned, landed, and pulled the gliders back into the air. The gliders delivered thirty-six urgent casualties to the 44th Evacuation Hospital in France, a feat that almost compared to modern aeromedical evacuation techniques.[52]

What to do with glider pilots once on the ground, however, presented a problem. In Normandy, Maj. Mark Alexander—the 505th PIR's executive officer—organized sixty pilots at the 82nd Airborne's division command post into a local security element defending the headquarters position, but this was far from standard procedure. Glider pilots in Holland sneaked off to Brussels and were charged with desertion, though the charges were later dropped. Ridgway thought that pilots should fly gliders and remain assigned to troop carrier units, while Gavin believed the pilots needed to be part of the airborne units and take ground combat training to be more useful. In Operation Varsity, glider pilots formed ad hoc units and performed well defending key terrain before being shuttled back across the Rhine.[53]

Aside from these critical moments, the glider was of secondary importance throughout the war. Gliders presented a tempting option to deliver massed combat power into an airhead but proved inadequate as soon as better airdrop capabilities and aircraft became available. The glider was imperfect, as it delivered larger—but not large enough—equipment, and the gliders themselves were often destroyed after a single combat deployment. However, the benefits of landing heavier antitank weapons, artillery, and transportation assets were evident, despite the loss of equipment and men. According to the postwar general board that studied their employment, airborne divisions must have "adequate artillery, adequate anti-tank means, adequate mobility, and adequate supply means for heavy and sustained fighting." The infantrymen riding inside gliders did not require specialized training like the paratroopers. The postwar General Board saw promise in the potential of improved gliders to "permit more rapid and orderly build-up on the ground," yet also—in 1945—considered the role that helicopters would play as "a better substitute" in the future. Helicopters were a natural evolution from gliders in their ability to deliver units together—rather than piecemeal via parachute—and the lack of a need for specialized training. Following Swing's example and the postwar board's recommendation, the army designated all units in airborne divisions as airborne rather than parachute or glider. The venerable old glider had one final training maneuver in 1949 but was officially made obsolete by the Joint Airborne Troop Board in 1953.[54]

During the war, airborne units earned an outsize reputation both inside and beyond the army. They performed well but did so while draining "a tremendous amount of talent at the cost of other units," according to Maxwell Taylor. "We were knee-deep in able young soldiers who could have been squad leaders, senior noncoms, platoon leaders, and company officers" in regular units. Regardless, the airborne's performance impressed Marshall so much that all four options given to Douglas MacArthur for his use as replacement commanders (in case of casualties or ineffectiveness) in the Pacific were airborne officers: Taylor, McAuliffe, Gavin, and Robert T. Frederick; the latter had commanded the First Airborne Task Force in Southern France. The postwar General Board noted that "in all operations, the airborne divisions displayed superior fighting qualities. Constituted entirely from volunteers, selected for initiative and aggressiveness, these units accomplished the most difficult of missions with distinction." When they were left on the line for too long, the performance of airborne divisions deteriorated, yet when bolstered by adequate attachments to approximate the composition and firepower of a standard infantry division, airborne units "were uniformly excellent."[55]

The 101st Airborne Division spearheaded American forces into Hitler's Alpine redoubt in Bavaria, quickly occupying the spiritual seat of National Socialist power in Berchtesgaden. Likewise, the 82nd Airborne Division was sent to Berlin for a most visible occupation duty, with

Gavin as the senior American in the city. While on occupation duty, Gavin created a special honor guard of highly decorated six-foot-tall combat veterans organized into one company that hosted General Patton. Patton called these men "the finest honor guard I have ever seen," and the 82nd's nickname as "America's Guard of Honor" was born. In Frankfurt, the 508th Parachute Infantry Regiment drew distinguished occupation duty—protecting General Eisenhower's home and headquarters until November 1946. In the Pacific, the first combat unit on Japanese soil was a battalion of the 188th Glider Infantry Regiment from the 11th Airborne Division.[56]

An airborne division was the natural choice to march in the victory parade scheduled for January 1946 in New York City. Initial army plans called for the inactivation of the 82nd in favor of the 101st, in order to include the Screaming Eagles' participation in the parade. However, Ridgway, Eisenhower, and the Army Ground Forces commander Jacob Devers wrote to recommend that the 82nd be retained, arguing that this was because of its longer, superior record, which included service in World War I. Plans changed after multiple strong endorsements in the press appeared in favor of keeping the 82nd on active duty. Gavin had weaponized his cozy relationship with the various journalists he encountered throughout the war, and their media blitz in support of the 82nd played a role in its retention on active duty and Gavin's overall stature. On January 12, 1946, Maj. Gen. James M. Gavin led his division down Fifth Avenue in a ticker-tape parade. Most of those marching were replacements, although many division veterans, including those wounded in action, were in attendance along the march route. That day's parade included the 555th Parachute Infantry Battalion, the army's only all-Black parachute unit—a critical visible step toward army-wide integration.[57]

Decentralization and individuality were essential traits in early operations in Sicily and Normandy. The experience of reinforcing the beachhead at Salerno and the Allied line in the Ardennes forest validated the requirement that airborne units maintain an expeditionary ethos and ability to improvise, which provided the impetus for later rapid-response forces. Meanwhile, solving tactical mobility and heavy-fire-support problems, alongside the complications of glider operations, reinforced the belief in adaptable thinking. The development of an airborne mindset was not only beneficial internally to airborne units but had external, lasting implications for the entire army. How airborne units fought and how they solved problems during World War II undergirded how their leaders approached issues later in their careers. Airborne officers learned valuable lessons about solving problems through organizational change, the efficacy of decentralized operations, the promise of tactical mobility, and the potential for airborne units to serve in a vital rapid-response capacity.

CHAPTER 3

The Airborne Way of War and Its Strategic Implications

> It was hard to break into the airborne syndrome group. After you get in it's great, but it's tough to break into that crowd.
>
> —Robert Haldane, 1985

When asked why he went to airborne school as a colonel at age forty-two, longtime cavalryman and future air assault acolyte Hamilton Howze joked that he "thought it was a good excuse to get away [from the Pentagon]" and that "they were all 'gung ho' characters. I just thought they were a good lot and I wanted to have some of the same experience they did." Howze wanted to join the club because he thought "they are among the best of our officers. I think there is a camaraderie among them." Three successive US Army chiefs of staff between 1953 and 1960 were airborne-qualified officers: Matthew B. Ridgway, Maxwell D. Taylor, and Lyman Lemnitzer. Like Howze, Lemnitzer was a late addition to the airborne club. Howze added significant value to the mafia, eventually bringing his penchant for mobility to the burgeoning air mobility faction. Lemnitzer was more of a careerist who decided to attempt parachute school at age fifty-one to take command of the 11th Airborne Division for a few months, check the block, and earn the all-important career marker of the 1950s army: jump wings. His decision indicates the airborne mafia's cultural and careerist draw.[1]

The airborne mafia's rise to the top was set against Cold War international competition with the Soviet Union. These officers rode their World War II reputations and personal connections to positions of increasing responsibility and impact during the Truman, Eisenhower, and Kennedy administrations as the United States set out its Cold War containment strategy and tweaked it over the following decades. Airborne units' World War II reputations fostered a new image of combat leadership to which many others aspired. The ascension of Ridgway, Taylor, and Lt. Gen. James M. Gavin to the highest levels was not preordained. They had each developed a reputation for trustworthy leadership and forthright thinking in the face of adversity. Most importantly, during fast-paced wartime conditions they had exhibited a flexibility that seemed to bode well in the new, rapidly changing atomic age. All three were forward-thinking West Point graduates known for their bravery, intelligence, and role in leading a brand-new dangerous experiment—the airborne division. All three were likewise committed to the idea that the ground soldier remained more important to modern warfare than machines, a stark contrast to a Department of Defense enamored with technology and atomic weapons to end wars quickly or prevent them altogether. Moreover, they drew others, like Howze and Lemnitzer, into their orbit.

The capacity to wage limited war was the focus of the army's struggle for political survival in the Eisenhower administration. Limited war at the time had multiple definitions. To the army during the 1950s, it meant any war below strategic nuclear exchange. Later, defense policy officials in the Kennedy and Johnson administrations considered it counterinsurgency in faraway places. Scholars of the time mostly thought of it like the Korean War—a war for limited political objectives. The army's ideas during the Eisenhower administration that flowered into flexible response were predicated on the need to have an alternative to massive retaliation for the defense of Europe, and this was first laid out in the 1954 edition of its capstone doctrinal manual, Field Manual 100–5, *Field Service Regulations: Operations*. Taylor was thinking specifically of deterring limited war in Europe, whereas Kennedy and his administration believed the capacity for limited war extended elsewhere—notably into Southeast Asia. Throughout the Eisenhower administration, the airborne mafia, by resisting air-atomic policies and advocating for a new flexible strategy better suited to meet the myriad nature of worldwide contingencies, played an essential role in preserving their service's relevance and independence in what amounted to a turf war inside the Pentagon over roles and missions. Flexible response was the army's and the airborne's suggested strategy, based on the belief in having more than just strategic weapons. During the Truman administration, the airborne mafia learned valuable strategic lessons about the importance of alliances, the role of atomic firepower, and the limits of American power overseas in a limited war. Their contributions in the Eisenhower era helped to firmly establish the airborne mafia as the predominant cohort of officers in the army. Finally, in Senator John F. Kennedy, the airborne mafia found a sympathetic ear in a politician running against Eisenhower.[2]

Learning Strategy in the Truman Era

Following World War II, the US Army experienced one of the most rapid drawdowns in its history. The army had 8,270,000 soldiers in uniform at the close of the war but only 590,000 in 1950 before the Korean War. This force was scattered worldwide yet at least nominally responsible for defending US allies against Soviet armed forces and other perceived threats. The army was also in shambles, with morale at an all-time low. It had trouble bringing in fresh recruits, leading to a reinstitution of a draft in 1948. The decline of the army became apparent when Communist forces invaded the southern half of the Korean Peninsula in June 1950. Meanwhile, atomic weapons brought a new and uncertain dynamic to future warfare.[3]

In the aftermath of Hiroshima, strategist Bernard Brodie wrote that until that point, "the chief purpose of our military establishment has been to win wars"; from then on, however, "its chief purpose must be to prevent them." President Harry S. Truman and his administration played a pivotal role in defining how the United States would contain the Soviet Union in the early years of the Cold War. In 1947, Truman signed the National Security Act into law, creating the Central Intelligence Agency and the US Air Force. In 1948, the world witnessed major events that heightened tensions between the superpowers—a Communist coup in Czechoslovakia, the Berlin blockade imposed by the Soviets, and the passage of the Marshall Plan for European recovery. In 1949, the United States joined the North Atlantic Treaty Organization (NATO) as a counter to Soviet power, while the Soviet Union detonated its first atomic bomb ahead of schedule, escalating the nuclear arms race.[4]

By the middle of 1950, the United States settled on a strategy of containment and deterrence that highlighted the gravity of the Communist threat while setting forth American objectives: stop the spread of communism, deter nuclear war, implant American capitalism on a global scale, and transplant American culture worldwide, all in accordance with the top-secret National Security Memorandum 68, better known as NSC 68. These goals, however, were an expensive undertaking that required the first large peacetime military force in American history. NSC 68 had a significant impact on the army and its role in the Cold War. The document helped the army advance its own goals to expand and modernize its forces. Specifically, it warned that the Soviets would threaten American interests using localized military actions worldwide. NSC 68 also outlined why an overreliance on strategic air power would weaken American diplomatic power and that only a larger ground force could deter Soviet threats. Jolted by the war on the Korean Peninsula to contain the Communist incursion south, the US military budget exceeded $48 billion by May 1951, a more than $33 billion increase from the pre–Korean War budget. During this period of immense contraction and uncertainty, the airborne mafia learned, grew, and experimented with new ideas before assuming positions of higher responsibility.[5]

The airborne mafia gained influence in the army's upper echelons during the postwar years. In late 1945, General Eisenhower appointed Matthew Ridgway to represent the US Army on the Military Committee of the United Nations in London. In this role, Ridgway advised the UN Security Council on military matters and served with Bernard Baruch on the Atomic Energy Committee. He became frustrated with debates among the "Big Five" UN members (US, UK, France, China, USSR) over military contributions and atomic weapons regulations. Ridgway saw Soviet disarmament proposals as a ploy to disarm the US and make the USSR the dominant global military power. During this time, Ridgway also chaired the Inter-American Defense Board, leading to the 1947 Rio Pact of mutual defense among Latin American states, which later modeled the North Atlantic Treaty.[6]

Taylor reported as superintendent of West Point on September 4, 1945, to rescue its slumping reputation. He was given a mandate from Eisenhower to reform the Academy. At West Point, he had an enormous role in shaping the school's curriculum, increasing humanities coursework to a full third of the cadets' curriculum. Leaning on his experience, on March 18, 1947, Taylor formed an airborne detachment that provided eighteen hours of instruction on "Airborne History, Airborne Equipment, Air Transportability, and the possible use of Airborne Troops in future wars." Taylor's airborne detachment grew into a requirement for all cadets to select airborne or ranger training upon graduation, and the detachment itself remained into the post-Cold War era. Even after airborne training was no longer required, it remained the most attended training school for West Point cadets. Taylor brought so many airborne officers into the faculty at West Point that it was referred to as the "101st Military Academy." In his next assignment, in Berlin, Taylor assumed command of all Allied troops in that Cold War outpost city. There he learned to think about more than just the military aspects of his command as he worked for both the Departments of State and Defense, in a proper politico-military position. He learned the value of combining all elements of national power under a single authority and toward a common goal. Moreover, Taylor learned that the value of deterrence and alliance, at least in Berlin, lay in the credibility of American forces.[7]

Gavin, meanwhile, commanded the 82nd Airborne Division until March 1948—three and a half years in command and six years with the division in total. Between 1946 and 1948, Gavin and his division were at the forefront of testing helicopter and atomic warfare techniques. In 1946, his division was assigned thirteen Bell helicopters, the first maneuver division granted that opportunity, and he found them "an air vehicle of great versatility." He viewed the mass parachute assaults in World War II as immediately outdated and began contemplating what the next war—one that might include battlefield atomic weapons—might look like. Gavin's first book, *Airborne Warfare*, argued that mobility, dispersion, speed, and stealth were instrumental to survivability on a future atomic battlefield. "The use, or threatened use, of atomic weapons has had one immediate effect on our nation's strategic and tactical thinking," he wrote, "the realization that dispersion must govern all operations of the future." Dispersion and decentralized leadership became necessary for the entire army—not just paratroops. He also believed that future roles for the army were predicated on the projection of military power through the air.[8]

After a brief assignment as Lt. Gen. Walton Walker's Fifth Army chief of staff, Gavin served as the senior army representative on the Weapon Systems Evaluation Group (WSEG) from 1949 to July 1951. The WSEG was a new organization, headed by Massachusetts Institute of Technology scientist Philip M. Morse and Lt. Gen. John Hull, to "provide rigorous, unprejudiced and independent analysis and evaluation of present and future weapons systems," according to Secretary of Defense James Forrestal. It launched in 1948 and comprised civilian scientists alongside officers from each service to ensure robust analysis of the "military payoff on weapons." During this period, Gavin attended the Nuclear Weapons School at Sandia Air Force Base outside Albuquerque, New Mexico, to learn the principles of nuclear fission and atomic bombs. His chief tasks involved leading a study of future airborne operations, developing tactical nuclear weapons, and devising bigger and faster air mobility systems, which spurred his conceptualization of "sky cavalry." Gavin's interest in tactical nuclear weapons was intensified after he participated in Project VISTA, a 1951–1952 study of atomic warfare against the Soviet Union in Europe. Among other conclusions, VISTA determined that tactical atomic weapons were ideal for offsetting Soviet numerical advantages.[9]

In September 1949, Ridgway became the deputy chief of staff for operations and administration at the Pentagon when Omar Bradley was elevated to chairman of the Joint Chiefs of Staff. Ridgway's role in the Pentagon would provide him with the necessary experience for his later role as army chief of staff. In his new position in the Washington bureaucracy, Ridgway was responsible for the day-to-day administration of a downsizing army. It was here that he first sensed "a growing feeling that in the armies of the future, the foot soldier would play only a very minor role" as he saw his colleagues enamored with "the erroneous belief that in the atomic missile, delivered by air, we had found the ultimate weapon." He watched as the secretary of defense, Louis A. Johnson, economized the army by reducing infantry regiments from three to two battalions, artillery battalions from three to two batteries, and by removing most medium tanks from infantry units. Johnson was trying to demonstrate his leadership abilities and favored the air force's way of war: he famously canceled the USS *United States* aircraft carrier in 1949—five days after its keel was laid—without consulting the secretary of the navy. His bold decision kicked off the "revolt of the admirals" and questions about which service should deliver a nuclear payload.[10]

When North Korean forces attacked across the 38th parallel on June 25, 1950, Ridgway was aghast at the lack of preparation. "The state of our Army in Japan at the outbreak of the Korean War was inexcusable," he wrote. As battle reports crossed his desk, he understood the situation better than most, and the bitter lessons of unpreparedness for Korea remained high in Ridgway's mind for the rest of his career. Following Walton Walker's death in a jeep accident, Ridgway was given command of the Eighth Army and an opportunity to shore up its performance. Back in his element while leading troops in Korea, Ridgway was always near the fighting and endeavored to raise morale through leading by example, as he learned by commanding in World War II. Ridgway's leadership salvaged the war in Korea, as within six weeks he turned a disintegrating, hollowed shell of a force into a victorious one. Maxwell Taylor described the turnaround in Korea as "the finest example of military leadership in this century." Ridgway was a soldier's soldier who dressed like a regular infantryman, replete with two hand grenades on his load-bearing equipment that earned him the nickname "Old Iron Tits." As he had done in the airborne, he preferred aggressive subordinate leaders who displayed personal leadership. In taking over a demoralized force, he believed that getting the men moving forward again was the only way to restore confidence and morale. He pushed commanders out of their command posts and relieved or rotated inefficient officers, fully understanding that soldiers needed leadership that shared hardship and led by example. Ridgway then succeeded MacArthur as commander in chief for the Far East, where he oversaw the broader war effort, began negotiations with the Communist forces, and worked to keep the war from expanding. For Ridgway, the biggest lesson from the Korea War lay in acknowledging the limits of American power, particularly air power, in achieving objectives, as the air campaign of "unrelenting pressure" to strike "devastating blows" designed to force North Korean concessions toward an armistice largely failed.[11]

Gavin went to Korea in October 1950 as part of the WSEG to evaluate weapons and tactics. There he witnessed how heavy fighting across arduous terrain presented multiple problems for the infantry—particularly in the two weeks it took forces that landed at Inchon to link up with forces pushing north from the Pusan perimeter. Like most army leaders, Gavin was appalled by the lack of preparedness, planning, and mobility exhibited by the units in Korea. In the rugged Korean Peninsula, Gavin began conceptualizing a helicopter-mounted force's vast possibilities in such terrain—something the United States Marine Corps had already started experimenting with in late 1951. In Korea, he realized that not only could a sky cavalry have been decisive, but so too could tactical atomic weapons if used on massing North Korean or Chinese forces. It was not in attacking population centers where Gavin saw the best use of the new weapon but as an extension of tactical firepower.[12]

After his tour in East Asia, Ridgway succeeded Eisenhower as supreme Allied commander for Allied forces in Europe in May 1952. Truman nominated Ridgway over Eisenhower's choice—General Alfred M. Gruenther—in part because of Ridgway's field command experience.

According to *Newsweek*, Ridgway's reputation as a fighting leader "was surpassed, among Army men on active duty, only by Eisenhower and General Omar N. Bradley." One of Ridgway's primary tasks in Europe was persuading the French to accept a rearmed West Germany within NATO. He also succeeded in reorganizing the command structure in Europe by simplifying the American portion, in order to provide unity of command over the three-pronged multinational force that stretched in a four-thousand-mile arc from Norway to Turkey. Despite remaining skeptical of European capabilities, Ridgway also developed in-depth contingency plans so that every subordinate commander knew what to do in the case of Soviet aggression.[13]

Gavin joined his mentor Ridgway in Europe, taking charge of VII Corps in December 1952. There Gavin honed many of the ideas he later used in developing concepts for fighting atomic war and air mobility. Leaning on his lessons leading the 82nd and responding to readiness issues he witnessed in Korea, he instituted practice alerts and other drills to ensure his subordinates were ready to meet a Soviet incursion at a moment's notice. Gavin became such an expert on nuclear weapons and warfare that he effectively served as the resident atomic warfare expert for the deputy supreme Allied commander, Field Marshal Sir Bernard Montgomery. Throughout his command of VII Corps, Gavin had his subordinate divisions thinking about fighting on atomic battlefields and testing concepts for dispersion. In October 1953, the army gave Lt. Gen. Anthony McAuliffe, Taylor's assistant during World War II, command of the Seventh Army. Like Gavin, McAuliffe expected decentralized leadership, encouraging subordinates to "think on their feet."[14]

After Berlin, Taylor returned to Washington in February 1951 for a brief tour of duty in the Pentagon. Taylor described his Pentagon work as "in effect preparation for participation in the Korean War as commander of the Eighth Army," though he could not have known that then. He served first as the assistant chief of staff, G-3 (operations and training). He was then promoted to deputy chief of staff for operations and administration. He later assumed command of the Eighth Army in Korea in 1953. Taylor's period in command, like that of his immediate predecessor, Gen. James Van Fleet, was marked by active defensive measures to preserve lives during ongoing peace negotiations. His tour of duty as commander in Korea included the unenviable tasks of implementing the terms of the June 1953 armistice, phasing out American forces and equipment, expanding the army of the Republic of Korea

(ROK), and fostering American relationships with South Korean leaders to rebuild the county's economy, as he had done in Berlin.[15]

While developing the ROK Army, Taylor experimented with an organizational structure he would later implement in the US Army. "I was convinced that our American triangular divisions, based on three large infantry regiments, was outmoded and regretted that it was being perpetuated in the new ROK Army," he wrote. In a study of the issue, X Corps commander Maj. Gen. Rueben Jenkins noted that infantry regiments had become too overburdened with administrative tasks and assets and proposed pooling resources at the division level while increasing firepower at the battalion level. Taylor proposed testing this within an ROK army division that included a headquarters with five subordinate maneuver elements rather than three. These streamlined battle groups were designed to operate dispersed yet assemble swiftly—as Taylor learned his airborne division could do in World War II. Ultimately, the South Koreans dismissed his experiment and opted to maintain a triangular structure. Taylor nonetheless remained a steadfast proponent of organizational change to best protect the force on the atomic battlefield. At posts around the world during the Truman administration, the three primary officers of the airborne mafia served in critical assignments in which they learned the value of allies but, most importantly, tinkered with tactical concepts that would come to undergird their strategic way of thinking as three- and four-star generals in Washington.[16]

"New Look" Insurgency in the Eisenhower Era

By 1954, each member of the airborne mafia had ascended to critical positions within the army. Ridgway was chief of staff, Taylor was in command of the Eighth Army in Korea, and Gavin was deputy chief of staff for plans and research. When Dwight D. Eisenhower assumed office in January 1953, he promised to take a "new look" at American national security strategy and find ways to reduce spending while continuing to contain the Soviet Union. By 1955, the air force budget was almost twice that of the army and remained so throughout the 1950s. Despite shepherding funds into the air force, the Eisenhower administration reduced defense expenditures by $6.1 billion from 1954 to 1956 while the air force grew by sixteen wings. Air force–delivered nuclear payloads were critical to Eisenhower's defense plans. Therefore, his administration viewed anything short of massive nuclear war—particularly limited local wars—as the responsibility of local actors and allies. Mutual security, alliances, and covert operations were thus paramount to American retrenchment behind atomic weapons, leaving little role for the army's large conventional formations. The air force and, to a lesser extent, the navy emphasized technology at the expense of large numbers of personnel. The army struggled to ensure its institutional survival amid a shifting international security environment and a fluctuating political environment. The creation of atomic weapons and strategic air power raised questions about whether ground forces were still essential and, if so, what their role should be.[17]

Eisenhower held a much broader view of war and military force than is often recognized. Branding him as a single-minded purveyor of atomic weapons is an oversimplification. The so-called New Look attempted to reduce spending through collective security, covert action, and broad propaganda efforts to showcase American supremacy as an economic and democratic system. This psychological warfare was accomplished through consistent messaging by the US Information Agency, the Atoms for Peace program, and the Open Skies Treaty. The latter two demonstrated America's willingness to cooperate, while the Atoms for Peace program sold the world on the peaceful application of nuclear technology. The army and the airborne mafia, however, articulated the Eisenhower administration's strategy in oversimplified terms explicitly to advocate its ideas for strategy and to argue for its institutional survival. They did so by suggesting that Eisenhower's grand strategy left the world vulnerable and that the army needed to fill a vital role through a more flexible approach.[18]

Meanwhile, the airborne mafia and armor officers developed competing conceptions of war. Tank commanders envisioned mechanized columns smashing through enemy armor, while the airborne viewed entire field armies flown deep into enemy rear areas and flanks. Neither would exist in total isolation, but the airborne view prevailed in the 1950s, primarily owing to lower costs for the army and because of the airborne mafia's leadership roles. Ridgway's experience in leading forces in two wars reinforced his belief in land power efficacy. He believed the army should maintain rapidly deployable, well-trained, and well-equipped units. The Korean War also showed him the limits of American power and the importance of ground forces. "In Korea," he later wrote, "we learned that air and naval power cannot win a war." He did not oppose airpower per se but rather the overemphasis on atomic bombing. As the New Look gutted army combat units, morale plunged, and leaders questioned their role. To Ridgway, the New Look's willingness to annihilate civilians in atomic warfare was morally bankrupt. He openly challenged policies leaning in that direction.[19]

Despite close ties during World War II, Eisenhower and Ridgway clashed over national security strategy during Ridgway's Pentagon tenure. Eisenhower entered office seeking to end the Korean War and curb spending. Through the New Look, the administration emphasized strategic bombing and continental air defense while deemphasizing ground forces, which had many leaders questioning the need for an army in a future war. To contain the Soviet Union without large, expensive conventional forces, the Eisenhower administration hoped to deter by the threat of nuclear weapons, supplemented by psychological operations, allies, and covert action. This approach emerged from Project Solarium, an exercise yielding proposals to contain communism without sparking general nuclear war—which Eisenhower saw as deterrable, given the immense risks. By threatening massive retaliation against aggression, the New Look compensated for military budget and manpower cuts with technology and nuclear primacy. These ideas clashed with Ridgway's views on sustaining versatile, resilient forces ready for atomic and nonatomic warfare alike, setting the stage for persistent conflict between the two men despite warm past relations.[20]

In his 1954 State of the Union address, Eisenhower denigrated the use of land forces in future war. He reiterated his position on nuclear weapons, stating that "the usefulness of these new weapons creates new relationships between men and materials. These new relationships permit economies in the use of men as we build forces suited to our situation in the world today. . . . The airpower of our Navy and air force is receiving heavy emphasis." Eisenhower also thought that the role of the army in a general nuclear war would be to restore order in American cities. He did not believe sending forces overseas was feasible or worthwhile and questioned Ridgway's intellect in implying that the general could not grasp the president's thinking. The president's message that the army would be relegated to inferior status infuriated Ridgway, who believed that the prospect of nuclear conflict meant that warfare would continue to require ground forces struggling to seize and defend terrain.[21]

Eisenhower had a much more holistic view of national defense. He presided over a robust system of allies and military aid, covert and psychological operations around the world, and understood that atomic weapons were not a panacea. Counterinsurgency was a focus during this time, particularly preventing Communist-inspired revolution. Much of the efforts during the Eisenhower administration were aimed at boosting internal security within friendly governments by providing money, arms, training, and advice. These military assistance programs also included funding for police forces and intelligence services in the 1954 Overseas Internal Security Program, which sent police officers to advise various friendly countries.[22]

To the army, however, the New Look was about institutional survival. In his confirmation hearing, Ridgway reiterated his belief in civilian control over the military. Officers, however—including himself—were duty-bound to present honest views without regard to partisan strife. Despite his professional outlook, he feuded with the secretary of defense (and former General Motors CEO) Charles E. Wilson throughout his two years as army chief of staff. Ridgway felt that Wilson came in with "absolute ignorance of the situation" in the Defense Department. Wilson asserted that "for years I thought what was good for our country was good for General Motors, and vice versa" and always tried to get the most "bang for the buck." Ridgway felt that Wilson treated the senior service chiefs like "a bunch of recalcitrant labor workers" and worse than he himself had ever been treated. Ridgway and Wilson's disagreements were evident to everyone in the Pentagon. Gen. Barksdale Hamlett said, "Wilson was out to get Ridgway; there is no doubt about it, and we knew it down on the staff." Budget and manpower totals continued dropping, something Ridgway believed "would [so] weaken the army that it could no longer carry out its missions." Ridgway further criticized the New Look's overreliance on machinery. "Because of the increasing complexity of land warfare and the resultant greater battlefield demands," he testified before the House Appropriations Subcommittee in 1955, "the individual soldier, far from receding in importance, is emerging ever more clearly as the ultimate key to victory." The army needed to be larger to fight and survive on the atomic battlefield.[23]

Ridgway believed that undue reliance on massive retaliation would handicap the nation's ability to defend itself or its allies against conventional and unconventional threats. His beliefs were validated during the encirclement of the French at Dien Bien Phu in 1954. When the French government appealed to the United States for assistance, the chairman of the Joint Chiefs of Staff, Admiral Arthur W. Radford, recommended a nuclear strike or massive conventional airstrikes. The proposed plan was known as Operation Vulture and called for large B-29 bomber raids, dropping fourteen hundred tons of ordnance and upward of three nuclear weapons on Viet Minh positions. Ridgway disagreed, seeing even conventional strikes as an escalatory step toward World War III. And that war would escalate to general nuclear war and become a conflict without a discernable political end state, something unacceptable to the astute Ridgway. The United States did not intervene, and the French garrison fell on May 7, 1954.[24]

After the French surrender at Dien Bien Phu and the creation of two Vietnams, American thought turned to supplanting the French in the containment of communism in the region. To that end, the administration considered an invasion of Hanoi through Haiphong harbor. That movement would have included the seizure of Hainan Island—a Chinese-held territory that the navy determined critical to placing its ships into the Gulf of Tonkin. Ridgway ordered his chief of plans for the army staff, Gavin, to develop potential courses of action. Gavin and his team concluded that American involvement in a rugged country lacking critical infrastructure would be costly. Their report proposed invading North Vietnam with seven American divisions organized in two corps, alongside a coalition of allies from France, the United Kingdom, Australia, New Zealand, the Philippines, Thailand, and other "associated states" (Vietnam, Cambodia, and Laos). Gavin also surmised that any invasion would greatly benefit the Soviet bloc, weaken the United States' abilities to defend Europe, and risk starting World War III. Such a war might cost the United States upward of $3.5 billion a year; and conflict in a roadless wilderness would entail a formidable engineering effort. The survey concluded that American intervention, besides its exorbitant cost, would allow "the Soviet Bloc to have succeeded in further dissipating US military power."[25]

Ridgway testified at a Senate hearing in May 1955 that downsizing US forces would alarm NATO allies. Yet as the army's commitments grew, its manpower shrank to "economize" the force. In a precursor to the Kennedy administration's "flexible response," Ridgway advanced four main points: (1) reliance on the capacity for "instant, massive atomic retaliation" leaves the free world unprepared to resist local aggression—as in Korea and Indochina; (2) if the United States did not possess an army capable of countering local aggression, it would invite local aggression and guarantee its success; (3) a strong, mobile, atomic-equipped army was just as essential as strategic bombing to deter and, if necessary, defeat aggression; and finally, (4) that an atomic army needed more, not fewer, men. Despite his deep concern over the adequacy of his force, Ridgway remained steadfast in his conviction that the army would make do with what it had, telling the House Armed Services Committee in 1955, "You may have complete confidence, gentlemen, that to the limit of its resources, the Army will continue to carry out its tasks and perform its assigned missions with unswerving fidelity, skill, and determination."[26]

In June 1955, Ridgway retired after only two years as chief of staff because, as he maintained, he had reached mandatory retirement age. But his scathing testimony at the 1955 Senate hearings did him no favors. Eisenhower declined to nominate him for a second two-year stint—despite the desire of the secretary of the army, Robert T. Stevens, to keep him in that position. He was essentially forced out, as a president could no longer maintain an army chief with whom he argued so often. In one particularly heated argument, Eisenhower accused Ridgway of "talking through his hat" for not realizing the reality of restoring order after an all-out atomic attack. On his way out, Ridgway sent Secretary Wilson and the president a letter voicing his concerns. He wrote that as then constructed, American military forces were "inadequate in strength and improperly proportioned" to meet large and small threats. Ridgway further argued that the United States' lead in nuclear capability failed to provide adequate bargaining power for the country's myriad crises during the mid-1950s. He also pointed out a significant fallacy in massive retaliation: as the Soviet Union gained atomic parity, Russian conventional force superiority again became a more dangerous threat.[27]

Gavin summed up Ridgway's service as chief of staff well. "Somehow, despite Secretary Wilson and the Chairman of the JCS [Radford]," Gavin wrote, "he managed to hold together our Army and to continue to ready it for the nuclear-missile space age despite a shrinking budget." Gavin noted that the budget was less of a problem than the "deception and duplicity of those with whom he had to work in the Department of Defense" during that era. Throughout a serial memoir published in the *Saturday Evening Post* that elicited comments from newspapers around the country, Ridgway continued to advocate for the army while explaining to a lay audience the fallacy of an overemphasis on air power and nuclear weapons. Ridgway's principal concerns were keeping the army relevant and out of politically ambiguous wars, a mission he accomplished through the careful application of dissent.[28]

Following command in Korea, Maxwell Taylor was nominated to succeed his former mentor, Ridgway, as the next army chief of staff. When President Eisenhower and Secretary of Defense Wilson interviewed him on February 24, 1955, they seemed more concerned with

Taylor's "willingness to accept and carry out the orders of civilian superiors" than with his strategic vision, reflecting their frustration with Ridgway's resistance to their policies. Taylor did not believe that nuclear weapons were a credible deterrent alone, but he agreed with Eisenhower on the importance of alliances over unilateral actions, favored arming allied military forces, desired a more centralized Department of Defense, and understood the importance of NATO. His understanding of NATO helped make the case that Taylor was the right choice to succeed Ridgway in 1955. Secretary of the Army Stevens had initially wanted Gruenther. Still, Eisenhower thought Gruenther could not be spared from Europe amid German rearmament and the alliance's various political, strategic, and diplomatic concerns.[29]

Taylor was sworn in as chief of staff on June 30, 1955. His deputy was his personal choice, Williston Palmer, who worked closely with Taylor as the VII Corps artillery commander in Normandy and briefly as commander of the 82nd Airborne Division in 1950, despite never

FIGURE 4. Gen. Matthew B. Ridgway, *left*, confers with his successor, Gen. Maxwell D. Taylor, June 23, 1955. Photo courtesy of the National Defense University Library, Special Collections, Archives and History. Ft. Lesley J. MacNair.

completing airborne school. Palmer had earlier commanded the Army Public Information School, and his expertise in public relations during an era when the army needed all the help it could get made him an intriguing choice. A 1955 memorandum summarized a meeting with the secretary of the army Wilbur Brucker, where Palmer admitted that the army was generally "rated last in comparison with the more glamorous Services" and that public opinion "forces them [Congress] to give priority to the Services which have a better grip on the public imagination. The Army needs to fight for its proper public opinion rating." Palmer was an able vice chief whom Taylor trusted with "complete authority to make the decisions on what you could call routine matters" to free Taylor to focus on the Joint Chiefs of Staff and big-picture items.[30]

Taylor was more critical of the Eisenhower administration's New Look policies than Ridgway had been. In 1956, he issued a memorandum to every command in the army outlining official positions on big-ticket issues to ensure his talking points reverberated throughout the service. Taylor believed in the likelihood of US involvement in limited war and that the Department of Defense's insistence on preparing for only general nuclear war was ill-suited to the realities of the modern world. He also understood that the army was in a minority position but endeavored to do what he could to fight for budget share. In a later statement to the rest of the Joint Chiefs of Staff, Taylor argued that the least likely type of war was "general war initiated by a Soviet surprise attack." To Taylor, the Korean War represented "another cogent argument against the emerging doctrine of massive retaliation" because nuclear weapons would prove useless against an entrenched enemy with a minimal industrial base, and the United States should instead focus on building capabilities to fight across a full spectrum of possibilities. In Taylor's estimation, a larger force structure was critical to deter and fight "local wars" around the globe.[31]

By 1956, the US Army's intellectual center at Fort Leavenworth, Kansas, devoted around half of its officer education curriculum, about six hundred hours, to scenarios involving atomic warfare. Yet despite insistence from leaders like Generals Taylor and Gavin about the likelihood of unconventional conflicts, the army made little progress in reorienting toward counterguerrilla operations. The army rejected proposals in 1954 to create dedicated counterguerrilla units, believing conventional forces could handle such threats. With an army consumed by atomic reorganization, it seemed better to arm and train allies to themselves fight insurgents under America's nuclear umbrella than commit US ground troops. However, in 1957, Taylor prodded the Command and General Staff College to draft plans to combat Communist insurgencies. Those plans called for providing threatened nations economic, financial, and technical aid to foster stability while building infrastructure to enable rapid US intervention. They also envisioned a force trained for counterguerrilla warfare and handling civil disturbances—the genesis of the Strategic Army Corps. This work recognized that defeating insurgencies required more than just military action, advocating an integrated approach to improve social and economic conditions.[32]

After Taylor assumed the role of army chief of staff, Gavin continued serving in the Pentagon as chief of research and development, overseeing the split of those responsibilities from the G-3 (operations) office. In this capacity, Gavin believed the army needed new capabilities in four key areas: a modernized cavalry for battlefield mobility, tactical nuclear weapons, enhanced strategic mobility, and pushing forcefully into missiles and space to match Soviet advances. Gavin argued that America required flexible military options to address "cold wars, limited wars, covert or overt aggression, general wars without nuclear weapons, or total wars with every weapon imaginable." He challenged the sufficiency of the threat of massive retaliation, noting that if the nation could not afford to fight limited wars, it could not afford to survive. Drawing on airborne units' adaptability in World War II, the "airborne mafia" offered an alternative vision of national security centered on flexible, adaptable options available to presidents and commanders rather than reliance solely on nuclear weapons.[33]

Taylor, for his part, understood that his job as chief of staff was to "make the best possible army out of the limited resources available to it, resources which I knew were dwindling." He believed that the army, if it was to obtain its share of the defense budget, needed to demonstrate an ability to deter war and fight an atomic one. "Like all other elements of our national defense programs," he said in an October 1955 speech, "the Army justified its existence primarily as a deterrent force to prevent wars." Later, in January 1956, when assessing the army's accomplishments during the previous year, he reiterated that its primary mission was to deter war, and its secondary mission was to win wars. To continue to do so required greater combat readiness "to deter the big war, but also the small war which may occur at any place about the world." Preparing an army for deterrence and war along a spectrum, with the dwindling resources of the New Look, required a creative mind, which

Taylor, with his background as a paratrooper leader and onetime West Point superintendent, definitely possessed.[34]

A peculiar phenomenon occurred during Gavin's tenure in the Pentagon. He often spent long days at the office, trying to work longer and harder while arguing more persistently than anyone else. His routine included Saturday "catch-up" sessions that grew into eighty to a hundred officers discussing policy and strategy in one of the Pentagon's auditoriums. These officers—referred to as the "Coordination Group," and unofficially sanctioned by Taylor—leaked information to the press and published dissenting opinions on national security. They were senior officers from around the Pentagon, usually colonels, frustrated with the Eisenhower administration's policies on massive retaliation. This group served as an unofficial political-intellectual planning staff. They fought an internal policy war to promote an agenda that sought to revitalize the army's roles and missions, using memorandums, articles, and leaks to other staff officers and sympathetic journalists.[35]

The chairman of the Joint Chiefs, Admiral Radford, insisted on total support for his views and shunned the sort of healthy debate and dissenting ideas that make for a better strategy. This also was an era of deep paranoia about Communist influence in the halls of government, thanks to Senator Joseph McCarthy. In such an atmosphere, the activities of the Coordination Group incensed Radford and other officials. One particularly pointed article that angered the chairman appeared in the *New York Times* on July 13, 1956, exposing a Radford plan for a reduction of nearly one-third in service personnel. "To some," Gavin said, "being against [the strategic primacy of] massive retaliation was seen as an attack against the Strategic Air Command and a possible sympathy for Communism." Ridgway, before he retired, had also, like Taylor, tacitly approved what the Coordination Group was doing. In instances when the actions of a particular individual displeased the administration, and officials asked Taylor to do something about it, he would "fire" the offender. This usually happened through his secretary of the General Staff, William C. Westmoreland, and the offender would often be sent to cushy assignment somewhere else around the army. But Taylor never took the blame.[36]

Later, Gavin's multiple congressional testimonies drew the ire of many in the Eisenhower administration, including Admiral Radford, to the point that Gavin found himself virtually ostracized. Exposing the realities of atomic warfare was a matter of national security to Gavin, so when asked, he told the truth. He believed in providing tough, realistic advice to his superiors. In 1956, Senator Stuart M. Symington (D-MO) held a series of closed-door hearings probing Eisenhower's defense policies. Symington was an ardent supporter of Ridgway and Gavin's ideas, and provided a platform to air their dismay to Congress. When asked by Senator James H. Duff (R-PA) during the Symington airpower hearings on May 25, 1956, about the effect of nuclear weapons exploding in Russia, Gavin deferred to an air force study but admitted that because of fallout, nuclear war with the Soviet Union would result in "several hundred million deaths," including allies in Japan or Western Europe, depending on wind direction. Congress had not realized just how devastating massive retaliation might be. When reports of that testimony surfaced in the *New York Times*, Gavin drew Radford's wrath for what Gavin felt was telling the truth, and thus became the scapegoat for the army's dissent.[37]

Gavin grew frustrated with the political atmosphere in Washington. He considered his boss, Taylor, less than forthright, and when reports publicized Gavin's testimony on nuclear war, Taylor never defended him. The two officers, so often linked, never got along. Gavin grew further disillusioned following a December 1957 hearing before the Senate Preparedness Subcommittee in which he testified that the Soviets beat the Americans in sending a satellite into space because the Joint Chiefs of Staff system did not function properly, with the service chiefs overburdened by serving as both leaders of their service and members of the Joint Chiefs. His testimony put him further at odds with the defense establishment. When Taylor informed Gavin on December 23, 1957, that he wanted to retain him in the Pentagon for another year, Gavin submitted his request for retirement within the hour. Gavin had wanted to go down to Fort Monroe in Virginia, assume leadership of Continental Army Command (CONARC), and do with sky cavalry what he had done with the airborne in the early 1940s, but to remain in the Pentagon at that point would have required him to betray his beliefs.[38]

Gavin's retirement played out on the front page of the *New York Times* in a series of articles from January 5 to April 1, 1958, the bulk coming in the second week of January. Remarking to Pulitzer Prize-winning journalist Hanson Baldwin on his attitude toward Pentagon politics, Gavin said, "I was taught as a cadet that a soldier's duty is to seek out danger. I did that in the war, and I was determined I was going to do it in Washington." He was urged to reconsider, offered to lead forces in Europe or at CONARC, with a promotion to four stars, and was also asked about running for the US Senate in his home state of

Pennsylvania. Ultimately, he remained steadfast in his decision, warning of another Korea-type war with an unprepared army, believing that he "had a choice of resigning, perjuring himself in testimony before Congress or being insubordinate." His influence on national security policy, however, was only just beginning.[39]

An Airborne Strategy for the Kennedy Era?

To deter Soviet aggression, the United States needed a flexible approach to national security, something that Ridgway, Taylor, and Gavin all argued while in uniform and in post-retirement books that offered scathing commentary on the New Look. Throughout his tenure as chief of staff, Taylor advocated for a more flexible strategy while loyally defending New Look budgets. Despite criticism, the Eisenhower administration did adjust course to provide more flexibility to meet national crises. American strategy after 1957 had begun to move toward more flexibility, as evidenced in the handling of the Beirut crisis in 1958 and the revised Basic National Security Policy of 1959—a clarification of existing policy. Nevertheless, opponents were convinced that it was not enough. So long as Eisenhower remained in office, his administration could not shake the impression among army officers that nuclear weapons would always take precedence.[40]

Taylor emphasized that the country had to be prepared to fight any type of war, from a general nuclear war to limited conventional war and "brushfire" wars of local aggression. While he defended the adequacy of the 1957 budget, he also used his day in Congress to outline his vision for deterrence at the local and strategic levels and the need to provide adequate means to fight limited wars in what was the first public acknowledgment of his thinking on the future policy of what would be called flexible response. Doing so required a robust mobile strategic reserve capable of rapidly reinforcing deployed forces or acting as an expeditionary force moving to trouble spots to deter and defeat local aggression. Mobility was the crucial component of his ideas but was not something the army could provide. It would have to be a borrowed mobility that necessitated more air force cooperation. During his four-year tenure as chief of staff, Taylor instituted two reorganization programs to best use the army's dwindling manpower while reasserting its importance in the New Look. He attempted to give the army atomic survivability and a new purpose in a dynamic international threat environment through his reforms.[41]

When John F. Kennedy won the presidency in 1960, he did so in part by using the same rhetoric that Ridgway, Taylor, and Gavin had used throughout the 1950s. Kennedy asserted that the New Look had caused damage to US military preparedness, reducing America's ability to influence the world. He campaigned hard on the purported "missile gap" as an attack on Eisenhower's policies. In an August 14, 1958, speech in the Senate, he quoted directly from Gavin's book that US "offensive and defensive capabilities will lag so far behind those of the Soviets as to place us in a position of great peril." Later, in June 1960, he argued for flexible capabilities when he declared that "we must regain the ability to intervene effectively and swiftly in any limited war anywhere in the world—augmenting, modernizing, and providing increased mobility and versatility for the conventional forces and weapons of the Army and Marine Corps." He likewise believed that nuclear retaliatory power alone was not enough. "In practice," Kennedy wrote, "our nuclear retaliatory power . . . cannot deter Communist aggression which is too limited to justify atomic war." Almost as soon as he entered office, his administration increased the defense budget by 15 percent and doubled the army's strategic reserve.[42]

After having waged insurgent resistance against Eisenhower, coping with the New Look, and facing massive retaliation in Washington, the airborne mafia believed the Kennedy administration's flexible response strategy was the alternative policy they had in mind. Carrying out a strategy of flexible response required building capacity for all levels of conflict. The airborne mafia played a critical role in developing the theories of flexible response adopted by the Kennedy administration as its national strategy. Taylor was recruited by the Kennedy administration in 1961, first to investigate the Bay of Pigs debacle and later to study counterinsurgency and activities that "fall short of outright war." He was brought back to active duty as "military representative of the president" upon completion of the Bay of Pigs investigation and was appointed chairman of the Joint Chiefs of Staff in October 1962, just in time for the Cuban Missile Crisis. He later served as the US ambassador to South Vietnam.[43]

While some believed Taylor had campaign ties with Kennedy, only Gavin was known to have spent time with Kennedy before the election. Gavin provided the future president with a copy of Taylor's book, recommended Taylor as an adviser, and pledged public and private support to the campaign and administration. Gavin maintained a cordial relationship with candidate Kennedy, talking at parties and even dining together. The general sent the candidate a steady stream of letters and suggestions relaying his ideas about the army, missiles, new weapons, defense, and foreign policy. Gavin responded quickly to Kennedy's questions and was rewarded with the position of grand marshal for the new president's inauguration, a position traditionally reserved for the chairman of the Joint Chiefs of Staff. Leading the parade in front of the Capitol was the 505th Parachute Infantry Regiment, the regiment Gavin had led into Sicily and which planted the first American flag on French soil in 1944. Kennedy also had Bill Walton, a journalist who jumped from the same airplane as Gavin over Normandy, as deputy grand marshal. Gavin was subsequently rewarded for his service to the Kennedy campaign with the ambassadorship to France during the tumultuous presidency of Charles de Gaulle.[44]

Kennedy and the airborne mafia understood the concept of what became known as "mutually assured destruction": that two (in this case) "rational" superpowers possessing scores of nuclear weapons would not use them and condemn their societies and the world to destruction. For Kennedy, Gavin, and Taylor, missile power would instead provide a shield under which limited war would reign supreme. While the Kennedy administration continued to fund and emphasize the American nuclear arsenal, the president and his secretary of defense, Robert S. McNamara, tried to place equal emphasis across the Department of Defense. Flexible response meant preparing to "react across the entire spectrum of possible challenge, for coping with anything from general atomic war to infiltrations and aggressions," in order to maintain an environment for the United States and its allies to prosper. It meant maintaining a flexible posture—something airborne leaders were used to in World War II—but on a strategic scale. Nuclear weapons remained an option, only to be used "as late as possible but as early as necessary." This represented a symmetrical approach to containing the Soviet Union, matching or countering the adversary's strengths in an attempt to maintain an international balance of power.[45]

The so-called missile gap continued to provide political ammunition to the Kennedy administration. Despite intelligence reports that the US enjoyed a missile advantage, Kennedy continued to increase and upgrade American strategic nuclear capabilities alongside his conventional force buildup. By mid-1964, the United States had doubled the number of Minuteman missiles the previous administration had ordered and added ten additional Polaris missile submarines. This constituted a 150 percent increase in nuclear weapons at that time. However, the administration also prioritized non-nuclear forces, and its policies tended toward decreasing the overall reliance on nuclear weapons to solve national problems. To this end it increased funding to the army, leaning heavily on many of Taylor and Gavin's ideas—transmitted through personal relationships between Kennedy and the two generals— that the world had reached a nuclear stalemate and the country needed a variety of forces available for all levels of conflict.[46]

Secretary of Defense McNamara concluded that the US defense structure—particularly in the army—left the United States vulnerable. The Berlin Crisis of 1961 only exacerbated these fears, pushing the administration to increase the number of conventional troops. As Taylor wrote, in extolling the inherently flexible nature of his service, "A B-52 bomber, an ICBM missile, or a Polaris submarine are good for use in general war and for little else. An Army division or a tactical air squadron has a use in any kind of war." After the Berlin Crisis, McNamara increased the number of active divisions from eleven to sixteen while pushing modernization efforts in all corners of the army. This force level was what McNamara estimated was required to handle a major war in Europe or Asia and a second, minor crisis elsewhere in the world. Army equipment began to receive priority, and during the Kennedy administration the army saw the development and modernization of the UH-1 and CH-47 helicopters, the M-60 tank, the M-113 armored personnel carrier, as well as the M-14 rifle, M-60 machine gun, and M-79 grenade launcher for the individual soldier. But, as the president stated in 1962, "these forces must be equipped and provisioned so they are ready to fight a limited war for a protracted period of time anywhere in the world," rather than just a general nuclear war.[47]

For Kennedy, a flexible response made sense because, as Gavin and Taylor had predicted, Soviet subversion in the Third World increased. Small, limited brushfire wars of local aggression would become the norm, and according to Taylor, the country needed a "balance of effort without undue reliance on an immediate resort to nuclear weapons to arrest the initial phases of aggression." Kennedy agreed. "Non-nuclear wars, and sub-limited or guerrilla warfare, have since 1945 constituted the most active and constant threat to Free World security. . . . Such conflicts do not justify and must not lead to a general attack," Kennedy told Congress in his first message regarding the defense budget, in March 1961. "Subversion and guerrilla warfare must rest on local populations and forces." He continued, "But given the great likelihood and seriousness of this threat, we must be prepared to make a substantial contribution in the form of strong, highly mobile forces trained in this type of warfare." This speech suggested an emphasis on supporting allies with counterinsurgent forces throughout the world.[48]

Crucial to carrying out the president's vision of a well-rounded military strategy was the capability for counterinsurgency warfare. Kennedy came into office wanting to prioritize efforts to combat Communist insurgent influence worldwide but did not have many specific ideas. His administration sought to understand the rationale of Eisenhower administration programs and to reinvent them to better serve revised priorities. One of the ways it did so was to elevate counterinsurgency as a higher priority within the Cold War and to associate it intellectually with modernization theory. This proved important following Khrushchev's promise in January 1961, just two weeks before Kennedy's inauguration, to support "wars of national liberation." On January 18, 1962, Kennedy signed National Security Action Memorandum (NSAM) 124, establishing the Special Group (Counterinsurgency) and brought in Taylor to lead it.[49]

The administration encouraged the Department of Defense to pour resources into Special Forces trained to understand irregular warfare's political, social, and economic aspects. Mandatory courses on counterinsurgency at the various war colleges and within the State Department, coupled with discussions of Mao Zedong's, Vo Nguyen Giap's, and Che Guevara's writings on guerrilla warfare, demonstrated the seriousness of administration efforts to focus on this sort of warfare. At Fort Leavenworth, instructional hours per student concerning the nuclear battlefield dropped from a high of 600 in the late 1950s to 53 in 1961 and 16 by 1966. Counterinsurgency-instruction hours there, meanwhile, ballooned from 35 to 222 between 1961 and 1969. Kennedy showed a personal interest in Special Forces training and equipment, including keeping a green beret—the symbol of the Special Forces— on his desk. During his administration, the number of Special Forces personnel at Fort Bragg, North Carolina, increased from fewer than a thousand to more than twelve thousand, and their training school now bears Kennedy's name.[50]

Late airborne convert Hamilton Howze chaired a Special Warfare Board in January 1962. That board evaluated and proposed the implementation of new instruction, doctrine, organizational structures, psychological operations, civic action (CA) capabilities, and specially designed individual and unit equipment, from uniforms to vehicles, aircraft, and radios. Howze also recommended assigning counterinsurgency as the focus for three divisions (including the 82nd and 101st). The board recommended a vast expansion of special warfare capabilities to meet the president and defense secretary's wishes. Likewise, Brig. Gen. William P. Yarborough, the inventor of the jump wings, was instrumental in reimagining and repurposing Special Forces as the archetype instrument for guerrilla warfare. After taking command of the Army Special Warfare Center at Fort Bragg in 1961, he immediately set out to "develop a new breed of man that could be sent out into the boondocks without supervision, who would continue to carry his nation's objectives in his mind." Like the visionary designers of the earliest parachute units, Special Forces leaders sought to build a new force comfortable in dispersed situations far from friendly support, in this case suitable to counterinsurgency.[51]

Likewise critical to Kennedy's activist foreign policy approach were global programs to cultivate American soft power, including the Alliance for Progress in Latin America, the new Agency for International Development, and the Peace Corps. These followed the "modernization theory" aiming to develop the Global South in America's image. The Alliance for Progress failed, largely owing to Walt Rostow's faulty assumptions about economic aid outside Western Europe. The Peace Corps combined ideas from Walter Reuther, Hubert Humphrey, Gavin, and others on harnessing young Americans' desire to make a difference. Reuther initially proposed the idea in 1950, and Gavin pitched it to candidate Kennedy in October 1960, providing a two-page memo one evening. Three days later, Kennedy used the idea in a speech. The name "Peace Corps," which Gavin suggested to Kennedy, came from a brainstorming session. More than just an altruistic way to help poor nations and give idealistic Americans foreign cultural experience, the Peace Corps was viewed by Kennedy as another Cold War weapon. In its first twenty-five years, the Corps sent over one hundred thousand Americans to serve in forty-four countries. It counterbalanced foreign policy debacles like the Bay of Pigs, engendering worldwide support for American ideals.[52]

"Flexible response" is necessarily a vague term; it assumes the necessity to act at all levels across a spectrum of warfare, a full range of available means to do so, and a response carefully calibrated to overall ends. While the Kennedy administration's defense buildup and increase of armor and mechanized forces might have signaled waning influence for the airborne mafia, it instead reflected an increase of their influence in the strategic and political sphere. The airborne mafia played an instrumental role in resisting Eisenhower-era policies, but their collective vision of strategy took root in the early 1960s. Nevertheless, under flexible response, Kennedy and McNamara prepared the army primarily for ground warfare, ostensibly in Europe, and emphasizing armored and mechanized units.[53]

The post–World War II moment was an awkward time to be a general officer in the United States Army. Airpower dominated strategy and the budget and left little room for the nation's ground forces. Atomic weapons threatened to eradicate humanity. War, being a political instrument, must be prosecuted toward some end, and general nuclear war never provided such an end. The airborne mafia never accepted the idea of the wholesale destruction of humankind or even the threat of such a thing and instead remained steadfast in their determination that war existed on a spectrum and required flexible options, not least of which included a prepared ground force. To that end, army leaders, particularly the airborne mafia, focused on making "the best possible army out of the limited resources available," just as they had done in the halcyon days of the airborne during World War II. Airborne officers began to see everything that flew over the combat zone as "airborne." One airborne-officer-turned-helicopter-proponent, Col. John "Jack" Tolson, even proposed a new "Airborne Corps with an entirely different connotation attached to the word 'Airborne' as to that used in the past. All personnel in this Corps will spend their careers in the aviation, rocket, and guided missile fields"—effectively lumping two of the significant developments of the 1950s under one term: airborne.[54]

The army during this period chose the paratrooper as its symbol of combat excellence, serving to exemplify its emphasis on the individual soldier. Such a choice, of course, was also the product of having the airborne mafia running every major facet of the army. Airborne operations accompanied most major exercises designed to test atomic warfare ideas. Two airborne divisions remained on active duty, and the number increased to three with Taylor's 1955 reactivation of the 101st Airborne. Its reactivation meant that the proportion of airborne divisions became three out of nineteen total divisions—nearly three times the proportion of airborne forces at the end of World War II. The impact of airborne officers on the army was so profound during this period that earning jump wings was considered a critical step for getting one's "card punched" on the way to the general officer ranks. This ought to come as no surprise, given that the three most influential army thinkers of this era were World War II airborne commanders.[55]

By the end of the 1950s, the leaders of the airborne mafia—Ridgway, Taylor, and Gavin—had all retired, Taylor's return notwithstanding. Still, their contributions, ideas, and actions had lasting consequences on army reorganization and civilian-military relations throughout subsequent administrations. The dissent of Ridgway, Taylor, and Gavin during the 1950s played a role in the Eisenhower administration's adoption of a revised national security policy in 1959 that reflected growing concerns about flexibility in responding to multiple crises. This did not reflect Taylor's vision, as he believed the current force structure was lacking but emphasized mutual security and a whole-of-government approach to meeting national security objectives. The rift between Eisenhower and army leadership also played a vital role in the Defense Reorganization Act of 1958, which removed the service secretaries from the chain of command and further isolated each service chief from the president's ear. By the time the less-experienced John F. Kennedy reached the Oval Office in 1961, the Department of Defense had virtually eliminated the service chiefs from their advisory roles.[56]

The airborne mafia was influenced by the same reasoning they used to support army employment of modern missiles, rockets, and helicopters—their values, beliefs, and norms from shared experiences commanding airborne divisions in World War II. While they had shared experiences, the triumvirate did not always see eye to eye, especially Taylor and Gavin. David Halberstam describes one White House incident where the young President Kennedy was excited to let Gavin know that Taylor had arrived, and the irascible Gavin gave the president "the coldest look imaginable" when Kennedy mentioned his longtime rival.[57] Nevertheless, learning to solve problems with organizational change, remaining forward-thinking, and being prepared to innovate while maintaining an expeditionary mindset paid dividends as the three led the army during the 1950s. Outside of their resistance to Eisenhower-era policy, their most lasting impact came in the atomic army, tactical air mobility, and strategic reach.

CHAPTER 4

The Airborne Influence on Atomic Warfare

> Fission in 1946 posed the greatest challenge to our military planners that we had ever known.
>
> —James M. Gavin, *War and Peace in the Space Age*, 1958

In terms of its cultural, intellectual, and doctrinal development, the atomic army might as well have been the airborne army. In fact, that is precisely what army leaders were trying to do during the 1950s—create an entire force structure light enough for complete air transportability yet robust and flexible enough to fight on an atomic battlefield. In testimony at a 1957 House Armed Services Committee, US Army chief of staff, Gen. Maxwell D. Taylor, stated, "All Army units must be trained for all-around combat in the same way we trained and fought our airborne divisions in World War II. . . . The infantry regiments . . . are administratively self-contained, air-transportable units organized essentially like the groups in the airborne division." This was Taylor's pledge to prepare ground combat forces for a potential atomic battlefield despite an administration that wished to relegate its primary land combat force to occupation duty in nuclear wastelands. Army thinkers, meanwhile, remained steadfast in their conviction that Soviet possession of atomic weapons meant the country would never again be able to mobilize and train after the onset of hostilities while the Allies bore the brunt of the first battles.[1]

The evolution of containment into the New Look reflected a belief that nuclear weapons had revolutionized warfare and that all branches of military service should reflect that reality. Because the air force and navy operated the primary delivery systems for strategic nuclear weapons required for massive retaliation, the army was relegated to inferior status. For the first time in history, the United States would maintain a large standing military force to prevent wars rather than raising one to fight them. Army leaders, however, maintained that if armed with atomic weapons, ground forces remained "indispensable to success in an era of guided missiles, intercontinental bombers, and atomic bombs." To maintain relevance and its share of the budget in the "air-atomic era," the army visualized a limited atomic land war against the Soviet Union that did not include atomic strikes on cities but rather exchanges of nuclear weapons at military targets. Priority was given "to the development and introduction into operating units of new weapons and techniques adaptable to the radically changed conditions imposed by the potential of nuclear warfare." This required fundamental changes in doctrine and force structure as the army realized it needed to join the other services in producing atomic capabilities. As Gavin put it years later, "You had to show you could live with nuclear weapons—either that or simply go out of business." In trying to justify its existence, regain prestige, and prepare to fight on the atomic battlefield, the army developed a doctrine and structure for such a fight that reflected the experiences of its leaders in that moment.[2]

The division structure that was developed during the 1950s mirrored the way the airborne generals' divisions functioned in combat: five subordinate maneuver units with limited artillery. How airborne divisions operated—dispersed, away from higher headquarters, and reliant on charismatic leaders, improvisation, and flexibility—was reflected in doctrinal concepts of the 1950s. Leaders realized that dispersal was critical to avoid presenting tempting targets to Soviet nuclear weapons. This required the entire army of the 1950s to function using the three tenets that drove airborne warfare in the 1940s: dispersion, mobility, and flexibility. In World War II, airborne forces became comfortable operating despite dispersion, whereas in hypothetical nuclear combat in the 1950s, dispersion became desirable. Taylor, Gavin, and Ridgway all came to similar conclusions that "smaller formations, quick decisions and improved communications, and the ability to disperse and assemble rapidly" were critical to survival in an atomic war, and the army attempted to transform the bulk of its force into mobile units that could strike, disperse, and reconsolidate at will. Rather than dispersing units with no command and control, this had to be *controlled* dispersion, as parachute units had been trained for. These characteristics are inherent qualities of the airborne subculture.[3]

Doctrine

Fighting in the atomic age required adjusting doctrine and unit organizational structures to deemphasize massed forces. Gavin's 1947 book, *Airborne Warfare*, encouraged many army theorists to envision a redesigned army that relied on mobility, dispersion, and low-yield atomic weapons. The army wanted to present a less enticing target for Soviet nuclear weapons yet be able to mass into larger formations when required for offensive action. The army needed to become a "hyper-mobile force" that relied on "nomadic tactics and greater mission-type delegation of authority to lower units." Gavin wrote that "the use, or threatened use, of atomic weapons, has had one immediate effect on our nation's strategic and tactical thinking—the realization that dispersion must govern all operations of the future." He was already testing ideas about dispersion in his 82nd Airborne Division, having his paratroopers assemble minutes before takeoff at their jump aircraft from dispersed assembly areas. The idea was to avoid presenting a tempting target for missiles or bombs, believing that Soviet commanders would be selective with when and where they used such expensive weapons. Later, after spending some time studying the effects of atomic technology as a member of the Weapon Systems Evaluation Group, Gavin concluded that "the A-bomb is an excellent tactical weapon" and that "the greater the explosive effect of the atomic bomb, the more effective it will be as a tactical weapon." The idea was that NATO and Soviet leaders would tacitly agree to limit atomic weapons to low-yield uses, only against military targets. By harnessing a tactical atomic capability, the army would be able to deter Soviet aggression because it could prevent escalation—at least that was the theory.[4]

Gavin was part of multiple studies on the efficacy of tactical atomic weapons while assigned to the Weapon Systems Evaluation Group (WSEG) in the Office of the Secretary of Defense. The most important was perhaps Project VISTA, conducted with the California Institute of Technology. Caltech had an impressive history of rocketry and nuclear physics research and thus served as a natural home for the project. Throughout 1951 and 1952, the WSEG and scientists from Caltech studied how to use American technology against the numerical superiority of Soviet forces. The project's report, issued in February 1952, argued for the tactical use of atomic weapons as a more effective way to stop a Soviet invasion of Western Europe than strategic bombing. VISTA argued "that nuclear power can be used to offset deficiencies in conventional arms." Air force leaders were incensed at this report, as it undermined the nature of the Strategic Air Command and massive retaliation and used language befitting flexible response instead. To Gavin, the point of the study was, in essence, to "bring the battle back to the battlefield," and he segued his efforts with VISTA and the WSEG into the rest of his career.[5]

While commanding the VII Corps in Germany in 1953, Gavin experimented with maneuvering his units on notional atomic battlefields. He concluded that the existing infantry and airborne divisions were insufficient for atomic war. Gavin believed the atomic battlefield would be much broader, deeper, and nonlinear. Any previous conception of a front would now double in length for each division. As Gavin said in one interview, "We are trying to discard old concepts of linear control of the battlefield for one of area control—a problem of controlled dispersion." This required isolated and dispersed battle groups, yet with more rather than less manpower. While in World War II an American division would hold a ten- to fifteen-mile front, Gavin anticipated that on the atomic battlefield, they might be asked to hold a front double in size and deeper than that any armed force had to contend with to that point. Gavin thus proposed a fundamental reorganization of how the army fights. With a much deeper atomic battlefield, linear defenses were no longer applicable, and units would have to fight as small, dispersed groups, presenting a less-tempting target for enemy atomic weapons.[6]

Soon after becoming chief of staff, Ridgway directed a radical review of the army's conduct of land warfare. Gavin led the study, and the review found logistics critical. "The key to atomic warfare tactically will be how to keep isolated units alive over long periods, still fighting when they are deeply encircled behind enemy lines," the group reported. The airborne mafia's shared experience with resupply by air in World War II reinforced their belief that aerial mobility provided the answer to the logistics, depth, and dispersion necessary in atomic warfare. Communications were also essential, and Gavin realized that battalions would need the amount of communications equipment he had used as a division commander in World War II. Missiles would function as improved artillery, providing the primary means of firepower to divisions, with an atomic capability if needed. As in airborne combat during World

War II, dispersion, flexibility, and mobility were the watchwords of the post-Korea army.[7]

The army's collective experience in Korea suggested that Communist forces' "combatant manpower" would be "the real tactical hurdle." How to make up for the manpower disadvantage seemed the critical question to answer. Gavin's vision of the nuclear battlefield provided a ready-made answer for how to stop the supposed Soviet steamroller. Missile-delivered nuclear weapons could function as a much-improved artillery system that, unlike an aerial-delivered atomic weapon, could function in all kinds of weather. The 1954 keystone manual, FM 100–5, *Operations*, instructed commanders to use atomic munitions as "additional firepower of larger magnitude to complement other available fire support for maneuvering forces." Army leaders also learned from Korea that future wars would have limited political objectives, rendering strategic atomic strikes unlikely. These leaders soon came to believe that limited nuclear warfare could be used to prevent a war from spiraling into general nuclear war. At the same time, low-yield nuclear weapons might deter Soviet aggression while offsetting their numerical advantage.[8]

Korea also reinforced that positional warfare would lead to a stalemate. Maneuver remained essential to victory. The depth of a future atomic battlefield would be enormous, perhaps hundreds of miles. Defense of such an area would require "scores, perhaps hundreds, of widely separated battle groups, relatively small in size but possessed of great mobility and tremendous firepower from conventional as well as atomic weapons." Moreover, as Gavin had learned commanding paratroopers in the previous decade, "the individual foot soldier in an atomic war must also possess far more initiative and self-reliance than in the past." Other officers also realized the value of mobility in atomic war. In the military journal *Armor*, Brig. Gen. Paul Disney wrote, "Mobility becomes the decisive factor in atomic warfare." Exercise Sage Brush, a multidivisional training exercise involving more than one hundred thousand soldiers conducted in Louisiana between October 31 and December 15, 1955, revealed the importance of battlefield mobility in a limited atomic war. Army forces would still seek to maneuver on a limited atomic battlefield to gain the upper hand and destroy the enemy army, it was assumed, as armies had done for millennia.[9]

Atomic firepower also enticed those looking for an easy alternative to sweeping maneuvers. Col. George C. Reinhardt and Lt. Col. William R. Kintner asserted that the frontal assault "would henceforth become the cheapest route after atomic weapons open the way." The idea was that dispersed battle groups would use nuclear firepower to blast a gap in the enemy front before beginning movement. These units would then mass and thrust through the gap created to exploit the enemy's rear area in a deep operation. This approach simplified coordination and emphasized exploitation rather than maneuver. The army at the time saw it as a revolutionary change in doctrine and set about to hone atomic skills among infantry formations at Fort Benning. However, the essentials were born in World War I, where many attacks included preparatory artillery barrages to reduce enemy defenses, followed by rapid advances to exploit the breach. In that regard, the emphasis on exploitation and attacking the enemy rear resembled German late-war *Stroßtruppen* tactics and combined-arms warfare in World War II. These leaders believed that low-yield battlefield nuclear weapons would accomplish what required millions of rounds of artillery ammunition in World War I France.[10]

The 1954 FM 100–5, *Operations*, prioritized division-level combined-arms operations. It was an update of the 1949 edition, with the addition of atomic warfare, which appears seventy-one times in the manual. Concepts of offensive operations remained unchanged and were predicated on ground offensives as in World War II and Korea, not on the latest thinking on atomic warfare, and the manual outright ignored general war. Despite its reference to atomic warfare, no change was noted in *how* the army should fight, only that atomic weapons provided more firepower alongside conventional munitions. "The integration of atomic weapons into tactical operations does not change tactical doctrine for the employment of firepower heretofore mentioned," the manual stated. It continued, "The planning and execution of offensive operations will continue to be based on the integration of fire and maneuver. Decisive results are obtained when a maneuvering force promptly exploits the destruction and psychological effects of atomic weapons." Officers placed importance on the atomic weapon to deliver the devastating firepower necessary to allow maneuver: "Speed, dispersion, flexibility—those keynotes of our whole era will take over the battlefield," wrote Col. Theodore C. Mataxis and Lt. Col. Seymour L. Goldberg in 1958. "Offensive tactics," they continued, "will be based on the atomic weapon; masses of atomic firepower will replace massed manpower in the attack." Firepower was beginning to outpace maneuver in both capabilities and emphasis in army thinking.[11]

Regarding defensive operations, dispersion continued to rule the day in depth and width between units. Like paratroopers operating behind the enemy lines in World War II, dispersion provided protection because small units were less-tempting targets. In the atomic age, this made them less vulnerable to missiles or other expensive weaponry, at least theoretically. No longer would units use assembly areas to mass before movement and attack. The difficulty lay in the theoretical instant massing of units in the moments after an atomic blast. As argued by atomic warfare proponents, the 1954 doctrine posited age-old defensive techniques, including positional and linear defenses, that were now obsolete. As outlined in FM 100-5, mobile defense seemed more useful insofar as it included forward "islands of resistance" that would slow attackers long enough so that they could be destroyed by mobile elements "at a favorable location." These islands of resistance were to be widely separated, using key terrain, and fight "an essentially independent battle." A further concept, the layered defense, envisioned a division area held by successive layers of battle groups, dispersed yet close enough to canalize and harass the enemy before counterattacking.[12]

Sustaining men and equipment on a dynamic battlefield provided another problem for atomic theorists. In 1958, Maj. Gen. Hamilton Howze wrote that an "increased use of air line of communication" was necessary to keep atomic units supplied on dispersed islands of resistance. Howze had been an armor and cavalry officer but by the mid-1950s had become a member of the airborne mafia. Frank Moorman, signal officer for the 82nd Airborne and later the XVIII Airborne Corps during World War II, recognized the need to create logistical units that reflected his airborne experience, "closely knit, well-trained, [and] hard-hitting," that operated with great flexibility for atomic warfare. Aerial resupply seemed promising in an atomic environment. Gavin thought twenty thousand airplanes for the army alone might not be enough and that for every one combat airplane, fifty more would be doing logistical work. Though the success of aerial resupply in Bastogne and the monumental achievement that was the Berlin airlift gave hope, the reality was that an airfield within a battle group's "island of resistance" earmarked for resupply would make a tempting target for Soviet nuclear munitions. Likewise, air-dropping supplies made transports easy targets for antiaircraft guns and surface-to-air missiles. Supply depots and ports were also determined to be probable enemy nuclear targets. The pentomic army never solved the sustainment issue before its abandonment in the early 1960s.[13]

One reason for the ultimate failure of the various developments of the 1950s was that the army never codified its atomic doctrine. In November 1951, the service published Field Manual 100-31, *Tactical Use of Atomic Weapons*, but these concepts never made it into its most important manual, FM 100-5, *Operations*. Army doctrine continued to emphasize infantry attacks at a deliberate pace, supported by tanks and heavy artillery. The army did not produce a new edition of FM 100-5 between 1954 and 1962, only supplements in 1956 and 1958 that did not address organizational changes. The Command and General Staff College submitted a draft manuscript to Continental Army Command (CONARC) in 1958, but it was returned with the comment that it needed to be fully revised.[14] This is despite Ridgway stating in an October 1954 interview that nuclear weapons compelled a change in tactics. Yet Ridgway maintained that fire and maneuver were still paramount, that the details of execution, distances to move, and explosive yield of firepower had changed.[15]

Instead of official doctrine, the army's unofficial doctrine—articles in the pages of professional journals such as *Military Review*, *Army*, *Army Combat Forces Journal*, alongside the writings and speeches of leaders—gave a good idea of what the army wanted its force to do on a nuclear battlefield. Even without official doctrine, the army rolled along, developed equipment for the atomic battlefield, and reorganized its divisions to meet its perceptions of limited nuclear war. The service was busy defining the parameters of limited nuclear exchanges. Taylor argued that small nuclear weapons would help the country avoid a general nuclear war, and units utilizing such weapons "had to fight independently, not linearly, and be expendable in the larger scheme if struck by a nuclear blast." Everything written about how to fight on an atomic battlefield was theoretical, tested only in exercises in the United States and Europe, with no means of real-world execution. Mataxis and Goldberg's book *Nuclear Tactics: Weapons and Firepower in the Pentomic Division, Battle Group, and Company* (1958) was still not official doctrine but called on leaders at all echelons to understand "atomic weapons" and to not treat them as a "specialized subject to be dealt with only by the staff officers or bulging-browed specialists."[16]

Taylor also believed that the army had to be dually capable—ready to fight in both conventional or atomic conflict—a nod to the immense flexibility required by both commanders and soldiers. Instead, units were never well prepared for anything other than nuclear war—even as the realization increased that the next war would not be nuclear. The secretary of the army agreed and declared that the army had achieved a dual-capable force ready for "all-out or limited war." Many officers were, naturally, displeased with his idea of a dual-capable force. One retired officer wrote that instead of providing a solution, the army only provided "those three stout words: dispersion, mobility, and flexibility." Some argued for the use of nuclear weapons in any conflict, while others argued for the opposite, and the debate played out in *Army* magazine.[17]

In devising its doctrine, the army also ignored the effects of radiation on its troops, expecting maneuvering forces to drive right through ground zero of a blast site if necessary. Dr. Douglas Lindsey, an army physician writing after Exercise Desert Rock VI, chided soldiers over their "bugaboo of radiation," which he believed stemmed from their "fear of sterility." Known for keeping a cigar in his mouth while commanding six MASH units in the Korean War, Lindsey was notoriously fast and loose regarding risk. The effects of radiation, he claimed, did not extend beyond fifteen hundred meters from ground zero. Meanwhile, maneuvering units had sat buttoned up and facing away from a thirty-kiloton atomic blast three thousand meters away during that exercise. Within minutes, they were advancing across the desert in pristine formation, only to turn away from ground zero at nine hundred meters when their radiation meters indicated dangerous levels of contamination. A later test in 1957 during Desert Rock VII featured a company of 82nd Airborne Division paratroopers lying on the ground forty-nine hundred yards from the detonation of an eleven-kiloton nuclear weapon on the top of a five-hundred-foot tower. The shock wave knocked the helmets off some of the troopers, who following the blast conducted an infiltration obstacle course to "test their ability to act efficiently in nuclear warfare." During Exercise Carte Blanche in Germany in June 1955, exercise umpires assessed that those nuclear attacks would have resulted in two million dead *before* radiation played a role. On the other hand, during Sage Brush, umpires failed to assess fallout casualties, which resulted in the indiscriminate use of notional atomic weapons.[18]

The army continued throughout the decade to refine how it wanted to fight on the nuclear battlefield. Americans have long preferred new technology and the overwhelming use of firepower on the battlefield; however, these atomic ideas were rooted in pure theory. Despite attempting to nest limited atomic war within the requirements of national strategy, leaders' conceptual thinking was insufficient for the intellectual demands of a hypothetical war. It would take equipment development and a reorganization of divisions to begin to put theory into practice. The doctrine was but one leg of the tripod of adaptation in the era.[19]

Technology

If doctrine is theoretical, then developing nuclear weapons for the army's inventory served as empirical evidence of the service's direction. Naturally, diving into the realm of weapon types whose only use had provided bomber enthusiasts an argument for the primacy of their arm was fraught with strife. The Key West Agreement of 1948 laid out the roles and missions of the services in the wake of the National Defense Act passed the previous year and the creation of the Air Force as its branch. In addition to outlining roles and missions concerning airlift, the Key West Agreement—amended in 1954—also gave the army a vague role in air defense. Thinking of atomic weapons as powerful artillery also fueled the development of guided and unguided surface-to-surface missiles. The focus on atomic missile and rocket technology, however, stymied growth and development in other areas as the army developed equipment to appease the secretary of defense and an atomic-enamored public to create a land force capable and relevant on an atomic battlefield. So, in 1950, the secretary of the army Frank Pace told the West Point graduating class that "the Army must depend to a great extent on intensive scientific research and development" to help offset enemy numerical advantages, setting the stage for a decade of targeted innovation. Shortly after becoming chief of staff, Taylor had learned of Secretary of Defense Wilson's priorities, ordering the army, in Taylor's words, to "substitute requests for 'newfangled' items with public appeal instead of the prosaic accouterments of the foot soldier."[20] Missiles and atomic weapons were the wave of the future and promised increased range, accuracy, reliability, and destructive capability.

Before it developed guided missiles, however, the army employed an atomic cannon. The M-65, 280mm artillery gun was a World War II design retrofitted for atomic ammunition beginning in 1947. Unveiled in October 1952 at Aberdeen Proving Ground, the gun fired a nuclear round for the first time in May 1953 and was featured in Eisenhower's inaugural parade. Nicknamed "Atomic Annie," the cannon could fire an eight-hundred-pound atomic shell almost fifteen miles, with roughly the same destructive power as the bomb dropped on Hiroshima. By October 1953, the first M-65-equipped unit arrived in Europe, the equipment floating up the Rhine River, visibly signaling to the Soviet Union the US Army's intention to defend Western Europe with any means necessary. The gun helped fuel the debates about roles and missions during the era; according to the chief of research and development, Maj. Gen. K. D. Nichols, "the Air Force did not set about the development of their [battlefield nuclear] capability until we started the 280-millimeter gun." Atomic Annie was the army's announcement that it intended to pursue atomic munitions.[21]

The gun, however, was obsolete almost as soon as it was fielded. It was slow, restricted to roads, and weighed eighty-three tons, which severely limited the number of bridges it could use. Like atomic doctrine, the gun was reminiscent of First World War ideas, specifically heavy rail-mounted artillery. The behemoth required not one but two tractors—one to push and one to pull the unwieldy cannon. Its paltry range was insufficient to strike deep into the enemy's rear. The range, speed, and mobility needed for atomic warfare were not found in the M-65. It was ill-suited to the fluid, amorphous atomic battlefield conjured up in the minds of army thinkers. Missiles provided a more enticing option.

When the army embarked on its missile programs, it worked in three areas: space, air defense, and tactical firepower. Surface-to-air missiles were viewed as defensive weapons, while surface-to-surface missiles found a primary purpose in offensive warfare, supporting ground forces. Throughout the decade, army leaders also envisioned placing a satellite in orbit to enhance communications platforms. When the Soviet Union launched *Sputnik*, Gavin was at Redstone Arsenal with the rocket scientist Werner von Braun, discussing Soviet rocket technology. Gavin predicted a launch within the next three weeks, only to learn of the Soviet accomplishment at dinner. When *Explorer I* achieved orbit using a modified army Jupiter-C rocket known as Juno-1 on February 1, 1958, it met raucous support from both houses of Congress, salvaging the nation's pride four months after *Sputnik*.[22]

The army's second major rocketry foray came in the form of surface-to-air missiles. While not especially relevant to limited atomic ground warfare, this foray was led by Gavin and centered on the Army's interpretation of the Key West Agreement, under which it was authorized "to organize, train, and equip air defense units." The army consequently developed the Nike and Hawk families of surface-to-air missiles, which frustrated the Air Force. The contentious issue was remedied by Secretary Wilson's 1956 amendment to the Key West Agreement, which clarified roles and missions regarding aviation. It gave the army responsibility for "land-based surface-to-air missile systems for point defense" while assigning the air force responsibility for "area defense." This meant that the army was responsible for defending specific sites, while the air force was responsible for maintaining a broader, networked air defense system. Air force leaders were convinced that the army was trying to take over the entirety of the aerial defense mission, including defending air bases, and would "wind up with an honest-to-goodness air defense mission." Air defense was integral to the army during this period but is outside the scope of this chapter's focus on land warfare.[23]

Gavin was among the army's most enthusiastic proponents of long-range missile technology. He believed that missiles could provide the fire support needed on deep airborne assaults, and when he became chief of research and development, missile technology became the top priority. Surface-to-surface long-range missiles offered more "bang for the buck" on the atomic battlefield of the future. Instead of expending thousands of rounds of ammunition, with the right guided missile a commander could destroy massing enemy reserves or blast a hole in a defensive front from hundreds of miles away, protecting his force at the same time. Missiles were also reliable and controlled by their service. Army leaders lacked faith in the ability of air force jets to support their ground operations. World War II and Korea had convinced army leaders that tactical close air support would be spotty in the next war, something all too obvious with the air force's institutional focus on the Strategic Air Command. Taylor spoke to commanders at Fort Bliss, Texas, in 1956, telling them that "we haven't had close effective tactical air support; we cannot expect to have it in the future. The high-performance air force planes are flying away from us; they have left the battlefield." Surface-to-surface missiles promised to remedy that problem.[24]

Missile advocates viewed them as the opening salvos of the next war. In 1947, Gavin wrote that "what was Pearl Harbor in 1941, followed in six months by an amphibious effort at Midway, will, in the future, be a missile barrage followed in six minutes or six hours by an airborne attack." He reiterated this in his 1958 book, stating that "the pattern of future attacks seemed to be emerging as a combined airborne-nuclear assault, using tactical weapons," something he felt received little reception in the air force despite the need for that service to deliver both arms and men. The air force considered that long-range missiles, if not a bomber replacement, were another form of strategic bombing and therefore the responsibility of the air component of the services, a view with which Wilson concurred in 1954.[25]

Missiles were mobile weapons that could complement field artillery, something that might maneuver with their amorphous battle groups to provide "a substantial improvement in ground combat effectiveness." In May 1947, the army fired its first surface-to-surface ballistic guided missile, Corporal E, at a range of 62.5 miles. By 1951, the army had already begun the development of a 450-mile ballistic missile. Throughout the period, the army developed three types of atomic missiles: surface-to-surface guided missiles such as the shorter-range Corporal or Redstone; intermediate-range ballistic missiles (IRBMs) like Jupiter; and shorter-range fire-and-forget projectiles for field commanders in tactical scenarios, like the Honest John, the Little John, and the Davy Crockett. These tactical missiles provided the most salient example of the army's attempts to adapt modern atomic technology to land warfare, to provide accurate and devastating supporting fire for ground commanders at all echelons. Gavin reiterated this view of missiles as improved firepower, writing, "For missiles are modern artillery and tactical nuclear warheads are modern conventional firepower."[26]

IRBMs became a matter of high priority for Gavin, who saw the technology as essential for obtaining adequate funds. He "personally recommended to General Ridgway in March of 1955 that the IRBM program be undertaken at a cost of $25 million." This led to the development of the Jupiter missile and the formation of the Army Ballistic Missile Agency in 1956. Gavin then ensured that the Jupiter IRBM be developed as a mobile and survivable piece of equipment to meet his vision of the atomic battlefield. The Jupiter was based on the Redstone missile, a 160-mile surface-to-surface ballistic missile developed between 1951 and 1953. The Jupiter was designed to be moved along highways, fire in all weather conditions, and even be hidden in highway and rail tunnels. The air force thought of missiles in terms of air bases and fixed launching sites requiring army defensive support, as opposed to the army's mobility requirement. While its initial range was just two hundred miles, the Jupiter later boasted a fifteen-hundred-mile capability, which frustrated the air force. Wilson restricted army guided missiles to a two-hundred-mile maximum range—anything over that was considered an intermediate-range ballistic missile that could attack beyond the immediate deep battlefield, and therefore air force responsibility. Previously, the air force claimed control over any missiles that could fly more than fifty miles into the enemy's rear. Two hundred miles, while still a limiting factor, was a significant increase for the army.[27]

Of course, Wilson's restriction did not sit well with the army. Gavin was never happy with Wilson's attitude and decision-making throughout the era. He wrote that Wilson "tended to deal with his Chiefs of Staff as though they were recalcitrant union bosses," while pondering if labor leader Walter Reuther might have been an efficient chief of staff. But for the army, it was a matter of an expanding battlefield. In defending his decision to pursue a fifteen-hundred-mile-range missile, Taylor reiterated the army's role in destroying enemy land forces. "The primary function of the Air Force is to destroy enemy air power and for the Navy to destroy enemy naval power," he testified in 1956; so "if you accept the fact that the Army exists to destroy hostile armies, then any missile which will destroy hostile ground forces should be available to the Army." The recently formed Association of the United States Army (AUSA) issued a scathing pamphlet critiquing Wilson's decision. The booklet argued that the range was immaterial—the army still envisioned tactical targets. "We define tactical targets as actual battlefield targets, troop formations, fortifications, and other man-made structures which immediately affect military operations on the land." This group of concerned retired officers accused Wilson of restricting the army from carrying out its primary mission. "Hence, there can be little logic in denying the army the tools it needs to carry out its primary mission." Jupiter was ordered into production in 1957, but in 1958 the air force assumed complete control of the program based on Wilson's 1956 decision.[28]

The army's short-range missile and rocket programs found much greater success. These were justifiable for use against military targets, and Gavin ensured they received funding priority. The army's first foray into short-range missiles was the Corporal. This liquid-fueled rocket could deliver an atomic or conventional warhead seventy-five miles with the help of an external guidance system that communicated commands to the missile via Doppler radio. First launched in 1947 yet not revealed to the public until 1951, the Corporal was operational by 1954 and was fielded in Europe in February 1955. However, the Corporal used liquid instead of solid fuel and therefore "lacked the responsiveness to provide truly effective support" to maneuvering units. The Corporal was considered a corps commander's weapon, able to affect the deep battlefield but not quite reliable enough for echelons below the corps level.[29]

Next, the army developed two types of surface-to-surface missiles: the longer-range Redstone, which became the basis for the Jupiter, and shorter-range, smaller weapons to provide atomic firepower to division commanders and below. The Redstone had a 240-mile range, which made the army the first force to field an IRBM before the program was handed over to the air force. Redstone also had an internal guidance system instead of the Corporal's external computer, which provided more reliability and an invulnerability to jamming efforts. The operator programmed the entire ballistic course into the computer, and the "internal course director" would compensate as needed to keep the missile on track. The Redstone was replaced in 1960 by the Pershing—a smaller, lighter, and more mobile solid-fuel missile with a 460-mile range. The new missile violated Wilson's range restriction but was developed under the guidance of the new defense secretary, Neil H. McElroy, to develop an improved solid-propellant ballistic missile.[30]

Short-range tactical weapons provided field commanders with flexible atomic firepower that fit the idea of mobile atomic battle. These smaller weapons with decreased yields fit Taylor's requirement to be suitable for use "in close proximity to our own troops." First developed was the Honest John, an unguided supersonic rocket with a fifteen-mile range that was later increased to thirty miles. While its range much improved upon that of the M-65 cannon, the missile was inaccurate, and its three-ton weight soon proved too heavy for the sort of battle-field mobility envisioned in the era—although the components could be slung beneath a helicopter separately. The Honest John, like the Corporal, was considered a corps-level weapon. Providing atomic weapons to division commanders and below required wholesale organizational change, beginning in 1956. As a result, in that year, the army began developing an even smaller, unguided, supersonic, nuclear-capable rocket, the Little John. Revealed in a public test in 1956 but not fielded until 1961, this rocket weighed only one ton but had a range of ten to twenty miles. It was light enough to be slung beneath a helicopter as a complete system and functioned as a division commander's weapon.[31]

The army's next evolution in atomic weapons placed the weapons' destructive power in the hands of battle group commanders. First named the Battle Group Atomic Delivery system, the M-28 Davy Crockett nuclear mortar weighed 185 pounds and had a range of just one and a quarter miles. The larger M-29 weighed 440 pounds and had a 2.5-mile range. The larger variant was designed to be carried by armored personnel carriers, to then be dismounted and fired from a tripod. The warhead had such an oblong shape that many soldiers dubbed it the "atomic watermelon." Yet it was the smallest nuclear warhead in the

Figure 5. Members of 1st Battle Group, 12th Infantry, sling-load an Honest John rocket to an H-21 helicopter for movement to a forward area. Spec. 4 G. L. Bowler, US Army photo, RG 111, NARA.

American arsenal, delivering an explosive yield of .01–.02 kilotons (10–20 tons). Its smooth bore, however, created massive accuracy problems, and the blast radius was rumored to exceed the maximum range of the weapon, which led many soldiers to call it a suicide weapon or the "widow-maker."[32]

To meet its concepts for waging limited atomic war while keeping up with the other services in a tight defense budget, the army focused on atomic weapons. Limited war was any war short of general nuclear war: one with limited objectives or limited means that could use atomic weapons. Focusing on atomic weapons left other more relevant programs earmarked for non-nuclear conflict with minimal funding. In a 1957 congressional inquiry, Gavin testified, "Because of the need to support the big ballistic missile program . . . we have had to cut back on the other things such as a new family of tanks." For example, in fiscal year 1957, the army spent more than 43 percent of its research and development budget on missiles and atomic weapons but only 4.5 percent on vehicles, 4.3 percent on artillery, and 4 percent on aircraft development. The army failed to update the cheap, reliable, and easy-to-maintain equipment the average draftee soldier used. Instead, it focused on flashy items sure to catch Secretary Wilson's or President Eisenhower's approval. Despite these disparities, Gavin remained steadfast in his conviction that "if the lessons of the decade since World War II mean anything, it is that this highly mobile mid-range missile with a tactical nuclear warhead should be developed as a matter of highest national priority" for use in limited warfare.[33]

Owing to the army's financial emphasis on missiles and atomic weapons, other useful programs developed slowly, if at all. Helicopters, tanks, and armored personnel carriers, ideal for moving foot soldiers about the nuclear battlefield and essential to survivability, made marginal progress during the era. Soldiers who served in World War II would have been comfortable using any of the infantry equipment available during the decade after the end of the Korean conflict. Expensive missile systems lacked the requisite targeting technology. Communications platforms that were needed to control units on a dispersed battlefield were on the drawing board and never came to fruition before the army changed direction. The helicopter offers a salient example of equipment development that might have been more useful in atomic warfare. Some officers saw the utility of helicopters, especially to move troops and equipment around the perceived nebulous, deep, atomic battlefield. But most planners during the era eschewed the role helicopters might play, preferring to think of mobility in fixed-wing strategic terms and relegating the burgeoning rotary-wing concept to medical evacuation, reconnaissance, and small-scale transport.[34]

The debacle over replacing the venerable old M-1 Garand rifle sums up nonatomic development in the era. Taylor wanted a versatile weapon to serve as a rifle, submachine gun, and automatic rifle, which delayed development. After it was fielded in 1961, the new M-14 proved inferior to the M-1 and inaccurate as an automatic rifle, and the army canceled production in 1963. During the 1950s, nearly all presidential hopefuls had a weapon system pet project that they touted to strengthen their campaigns—most were high-technology or nuclear-related like bombers and missiles—never the needs of the foot soldier. When one staffer asked a young defense intellectual named Daniel Ellsberg what the Kennedy campaign should use, he replied tersely, "How about the infantryman?"[35]

Developing missile technology was expensive. Yet the army pursued this equipment at the expense of other items in the hopes of improving its public image and maintaining relevance in an age where many policymakers were questioning its very existence. By showing it could develop and use atomic weapons, the army kept pace with the other services and played an enormous role in the space race. On January 31, 1958, the nation's first successful satellite rocketed into orbit atop a modified Jupiter missile, thanks to the work of scientists at the Army Ballistic Missile Agency. The rejection of increases in armor in favor of providing devastating (atomic) firepower to units reflected the shared belief held by the airborne mafia in the ability of well-trained groups of infantrymen to move quickly, mass available firepower, and fight dispersed on any battlefield on which they found themselves. Moreover, missiles and rockets impressed the public. The army needed the proper organizational structure to take advantage of them.[36]

The Pentomic Division

As the army continued to think about fighting on the nuclear battlefield, it became clear that it needed to reorganize the infantry division to meet emerging concepts in doctrine and take advantage of advances in tactical nuclear weapons. Gavin believed the new division should be "an amorphous biological cell" capable of responding to any situation by taking on whatever form necessary. For Gavin, this meant dissolving the existing organization "down to the size of units you are not afraid of losing to one blast" and that the infantry division needed to mimic the mobile combat commands of the armored division. The battalion—renamed the battle group—was the ideal unit size capable of sustained combat yet expendable in a nuclear blast. Each battle group would have the assets needed to deploy and disperse in all directions. It was intended to be more sustainable thanks to increased reconnaissance, signal, maintenance, and medical assets in a battle group headquarters and service company.[37]

In April 1954, Ridgway directed Army Field Forces to create a new division predicated on mobility and flexibility. Eisenhower believed that emerging technologies would permit "economies in the use of men as we build forces suited to our situation in the world today" and allow for a smaller army. Ridgway's guidance asked Army Field Forces to meet several critical objectives. They needed greater combat manpower ratios, greater combat-to-support unit ratios, and more flexibility and mobility in combat units in order to maximize new technology and improve the ability of the army to sustain land combat. And they needed to develop tactical doctrine to support these changes and reorganize all units by January 1, 1956. Ridgway's ideas for a ground combat division envisioned larger, heavier divisions able to absorb more losses. In Ridgway's conceptualization, decentralization on the battlefield through greater dispersion was only possible with leaders capable of and comfortable with making plans and decisions with minimal guidance—empowered subordinates. It reinforced his thesis that human beings were the dominant factor in war.[38]

Project Binnacle, a US Army War College effort, studied the effects of nuclear weapons on the army in the next decade. In addition to arguing that mutual annihilation through massive retaliation would keep wars limited, the study concluded that the army must be "designed to locate, maneuver, and fix enemy forces and then blast such forces with nuclear fire." The study also proposed a reorganization of the army into 12,000-man divisions of multiple 1,000-soldier combat teams capable of air and ground mobility and armed with atomic weapons in addition to their conventional complement. Project Binnacle served as the basis for the army's next foray into divisional reorganization conducted by Gen. John E. Dahlquist's Office of the Chief of Army Field Forces.[39]

The resulting plan, which filtered from Army Field Forces through the Command and General Staff College at Fort Leavenworth, Kansas, in the fall of 1954, was titled Atomic Field Army-1 (ATFA-1). The study results hinted that the cellular battlefield envisioned by Gavin was likely but required new tactics and unit structures. Improvements in communications equipment and experience in World War II hinted that a division could control up to five maneuver elements rather than just three regiments. In proposing a 4,000-man reduction for each infantry division, ATFA-1's strength stood at 13,500 officers and men. It appeared to be the perfect nexus of flexibility, mobility, and firepower. Announced in the February 12, 1955, issue of the *Army–Navy–Air Force Register*, the new organization, as Gavin reiterated, would be ready for conventional or atomic war, and "he expected non-atomic 'peripheral' war the more likely type as both East and West approach a state of 'thermo-nuclear plenty.'" The article emphasized "communications, intelligence, firepower and above all mobility to keep from concentrating in one spot."[40]

The ATFA-1 division would have had seven infantry battalions alongside tank, engineer, and other support elements. Division artillery was weakened, containing only one 4.2-inch mortar battalion and two 105mm howitzer battalions. It was a much heavier division than what was ultimately decided. The division lacked antitank and artillery support, and the personnel reductions left staff at all echelons unable to perform essential support functions, as many administrative or logistical matters were to be handled by a support command. Aviation and reconnaissance assets were consolidated in companies within the division headquarters battalion. The Infantry School was concerned about the lack of a regimental command level and proposed using a "combat command" in the style of armored divisions. ATFA-1 proved untenable and never came to fruition after February 1955 tests by the 3rd Infantry and 1st Armored Divisions in Exercises Follow Me and Blue Bolt showcased its issues in command-and-control capability. To solve the problems encountered, the infantry division needed an increase in its staff and an additional infantry battalion.[41]

Later, in 1955, the same divisions tested a retooled version of the structure during Exercise Sage Brush at what is now Fort Johnson, Louisiana, and determined that the ATFA concept "did not appear to have any greater mobility, or any greater capability to conduct successful atomic-tactical operations than did the conventionally organized forces." Sage Brush was the largest maneuver exercise since the 1941 Louisiana Maneuvers. It was designed to test the new organizational structure amid simulated atomic strikes. Observers also noted that the ATFA organization showed "no marked improvement" over the previous triangular formation. To further complicate matters, Army Field Forces was reorganized into the Continental Army Command (CONARC) to provide a better command element between the chief of staff and the six Continental Armies arrayed around the country. CONARC then reinstated a triangular division structure, abandoning the ATFA-1 concept.[42]

After taking over as army chief of staff, Maxwell Taylor rejected the ATFA-1 concept because it did not create smaller, simpler division structures with tactical atomic weapons. Later, in 1956, Taylor selected a different study for implementation that came from an earlier rejected proposal. Titled "Doctrinal and Organizational Concepts for Atomic-Nonatomic Army during the Period 1960–1970," but referred to as the PENTANA study, this model called for fully air-transportable divisions based on five self-sufficient battle groups, each with its artillery. Each division would have 8,600 men and eliminate the regimental level of command—a more than 50 percent reduction in personnel and the number of aircraft required to move it. The PENTANA study and Taylor's desire to implement it met with staunch dissent from the army's

Command and General Staff College (CGSC) at Fort Leavenworth. Namely, two successive commandants, Maj. Gen. Harry I. Hodes and his successor, Maj. Gen. Garrison H. Davidson, viewed it as fostering an officer corps too narrowly focused on atomic warfare. By far Taylor's most ambitious project, PENTANA was approved on June 1, 1956. And in true Taylor style, he replaced Davidson with a sycophant, Maj. Gen. Lionel C. McCarr, in July 1956, and promptly revised the entire CGSC curriculum for atomic combat.[43]

PENTANA was another, lighter response to Ridgway's November 1954 directive to find a new organization capable of atomic and conventional warfare that applied to sustained ground combat on the Eurasian landmass. The new structure offered more flexibility in its five-battle-group structure. Like the ATFA-1 division, however, PENTANA offered meager artillery support despite new Honest John nuclear rockets. It also lacked tanks, antiaircraft artillery, engineer, and reconnaissance assets. Before approval of PENTANA, Taylor directed a reorganization effort for airborne divisions using a modified version, which incorporated portions of both the PENTANA and the ATFA-1 studies. This proposed airborne division would have five battle groups, four infantry companies, a 4.2-inch mortar battery, and requisite support assets. The division included a divisional support group, a command-and-control battalion, a signal battalion, and a small engineer battalion with airstrip construction resources—critical for airfield operations. Division artillery assets were slim—three 105mm howitzer batteries and one nuclear weapons battery with Honest John rockets. The Little John and Davy Crockett were not yet available, while 155mm howitzers were eschewed due to their lack of air transportability.[44]

Taylor endorsed this change in February 1956 but ordered the addition of a fifth infantry company and an increase in 105mm batteries from three to five. ROTAD (Reorganization of the Airborne Division) was approved by Taylor on August 10, 1956, as the airborne version of the concept. The airborne division was ideal to test the new five-sided divisional structure for atomic battlefields. After Taylor reactivated the 101st Airborne Division in 1955, the army again had three airborne divisions on active duty. The 101st moved to Fort Campbell in 1956 to test the new concept and replace the 82nd Airborne Division as the Western Hemisphere Reserve. The 101st tested the idea of an "amoeba"-like fighting concept while stressing the need for more helicopters to improve mobility. Those initial experiments preceded the massive test known as Jump Light in August of that year, which kicked off the five-year experiment with what was to be called the pentomic structure. The division's connection to the chief of staff, its World War II record, and public recognition made it the logical choice to spearhead this effort.[45]

The new airborne division organization also meant it was completely air transportable—a capability it had lost in the decade since World War II. Air transportability was a critical requirement for the army in the atomic age. While leaders strove to improve the transportability of all units, the airborne division alone met the army's stiff criteria for strategic mobility. But to do so, it lacked tanks, armored personnel carriers, and conventional fire support larger than 105mm howitzers. The choice to go without robust fire support indicates two realities—the need for air transportability and the belief that any war would include the use of tactical atomic weapons. Regardless, Taylor had all army divisions reorganized along this basic structure to survive in omnidirectional atomic or conventional combat. Initial plans for the new "pentomic division" now called for reorganization to be complete by the end of May 1958, and after a few minor delays, reorganization of the army's fifteen combat divisions, one brigade, and several separate battle groups was completed by June 30, 1958.[46]

The term "pentomic" was an attempt to sell the army to the public. Taylor admitted it was a "Madison Avenue adjective" that referred to its pentagonal structure and atomic purpose, designed to cater to Secretary Wilson's desire for "public appeal." Nevertheless, the five-sided structure was necessary for fighting on Gavin's nebulous atomic battlefield. Each battle group was to fight alone in all directions, if necessary, creating islands that forced the enemy to present themselves as targets for the division's internal tactical nuclear weapons. Theoretically, the dispersion required depended on the adversary's willingness to expend its nuclear weapons on smaller targets. The new configuration meant increased air transportability and battlefield mobility for all divisions. Every division was to be a worldwide air-deployable formation. All five battle groups could be lifted with just over 400 sorties of aircraft, a significant reduction from the 620 sorties needed for the assault elements of the previous 17,000-man triangular infantry division.[47]

Midway through the restructuring effort, Maj. Gen. T. L. Sherburne of the 101st wrote that his new division was so transportable, so mobile, that it needed "less than half the planes required for a conventionally manned and equipped airborne division" and was fast becoming the army's first airborne "True Ready Force." He boasted that it would be ready to "move out, completely ready, *in a few hours*" and strike hard whenever and wherever the country asked. As the word "mobility" continued to drive the army, the pentomic division promised fluid concentration and dispersion as required to fight on the atomic battlefield. The idea, in practice, was to concentrate, strike, and then disperse as needed. However, without adequate equipment, the quick tempo promised to tax even the army's most skilled commanders. "Concentrate to fight—disperse to live," one army officer once said.[48]

The pentomic division reflected airborne leaders' shared experiences. The division was predicated on an increased span of control in fluid, ambiguous situations, as Ridgway, Taylor, and Gavin experienced in World War II. During that war, the 101st and 82nd Airborne Divisions fought as ever-changing divisional-size task forces, adding or subtracting regiments depending on the situation. During the invasion of Normandy, for example, each airborne division had four infantry regiments under its command. During the Battle of the Bulge, the 101st Airborne Division managed to control five separate maneuver units, including two armored combat commands. The concept also has roots in Taylor's prewar experience as commander of the 12th Field Artillery Battalion, where he commanded five separate batteries. Taylor had further experimented with a division having five subordinate elements in the Republic of Korea army while in command of the US Eighth Army in 1953. Despite the influence of the airborne mafia, one cultural tenet was not fully disseminated to the army of the 1950s—effective decentralized leadership. Instead, the army was characterized as overcontrolled and oversupervised. This resulted from rigid career management systems imposed after World War II that encouraged a careerist mentality among officers.[49]

The pentomic structure was not well received by most of the army. Units in Germany found it wanting, yet CONARC evaluated it positively and saw it as well suited for conventional or atomic warfare. A 1963 army study noted that it lacked flexibility and confined the army to one type of warfare: limited atomic warfare in Europe. Tactical mobility was inefficient, and the pentomic division had less conventional firepower than the standard infantry divisions in World War II and Korea. The Seventh Army concluded that a pentomic division had insufficient conventional *and* atomic firepower—which in effect removed the strength of American artillery since World War II, namely its centralized command and fire-direction control. Gavin commented that it was unfit for combat, "was really a mess," and needed more artillery, because "if you've ever had to use artillery, you knew it was grossly lacking in artillery." He saw Taylor's swift approval of the pentomic division as typical—little analysis but always wanting to please politicians. The pentomic rebrand on Madison Avenue also points to Taylor's efforts to showcase the army as new and exciting in the atomic age. Nevertheless, the pentomic battle group's size fell well short of Gavin's ideal of "two or three thousand men" per battle group.[50]

The division had further problems; the span of control for commanders was too great because the necessary communications equipment did not yet exist to facilitate a commander's ability to control five subordinate units at the distances envisioned on an atomic battlefield. The army knew it could not provide the equipment necessary for full attainment of combat effectiveness. In September 1958 the Seventh Army commander, Lt. Gen. Clyde D. Eddleman—a career infantryman who served on the Sixth Army staff in the Pacific during World War II—recognized that the army's "current capabilities with weapons and equipment on hand do not permit us to enjoy the full benefits of the new concept." One officer wrote in *Army* magazine, "Despite a greatly revised organization and tactical doctrine, combat units, as usual, are trying to do with the same old equipment until the new gear arrives." The pentomic division, while well suited to quick-hitting airborne assaults and short ground operations, appeared insufficient for low- and mid-intensity conflicts. Austerity in personnel, firepower, and logistical support required additional assets if the division was to be committed to a long-term conflict.[51]

The pentomic division was supposed to demonstrate that the army was a forward-thinking force worthy of a meaningful place in the New Look military. Instead, it revealed that the army overestimated its ability to create a highly mobile, well-supplied force. Under the circumstances, its replacement was being considered as early as January 1959 under the Modern Mobile Army 1965–1970 (MOMAR I) study initiated by the CONARC commander, Gen. Bruce Clarke. MOMAR I reflected Clarke's experience as an armor officer and commander of the Seventh Army from 1956 to 1958 before taking over CONARC in 1959. He believed in concentrated rapid armor attack—fast and in-depth—and MOMAR I provided that. His division structure emphasized mechanized and armored vehicles—a heavier approach signifying the waning influence of the light- and air-minded airborne mafia.[52]

Eddleman succeeded Clarke as the commander of the Seventh Army. While in Germany, the two had worked closely with the reconstituted

German Bundeswehr, observed their new Panzergrenadier divisions, and realized the limitations of the pentomic division. The Germans had come to many of the same conclusions as the Americans: tactical atomic weapons would require mobility, dispersion, and flexibility and were necessary to offset Soviet advantages. The Germans likewise prized the flexibility of tailoring divisions with units as needed and therefore preferred a brigade structure with tailorable units. In 1959, the Germans unveiled a new division based on three tactical brigade headquarters that were self-contained. Battalions were reduced in size to streamline their capabilities. Each division had twelve battalions rather than nine. All infantry became armored infantry, and brigades of several maneuver units were capable of independent action. The division headquarters essentially became an administrative formation. The Bundeswehr's brigade structure became the standard for all NATO divisions in the late 1950s to achieve greater compatibility across the alliance. The NATO Standing Group concluded that the division concept adopted by the reconstituted German army was to be the standard across the alliance, though the United States had yet to settle on its new concept.[53]

MOMAR I was designed explicitly for war against the Soviet Union in Central Europe. Eddleman wanted the army to develop a tailorable division structure like that of the Germans, which could be used in various hot spots around the globe. As commander in Europe, he had combined battle groups into quasi-brigades. In one of his first acts as vice chief of staff, Eddleman instructed the CONARC commander, Gen. Herbert B. Powell, to devise plans for infantry, armored, and mechanized divisions. They needed flexibility in composition. He wanted a flexible brigade level of command based on the examples set by NATO Central Army Group and Bundeswehr divisions. The new study was submitted less than three months later.[54]

Initially titled MOMAR II, the new division was supposed to provide flexibility that bridged the capability gap between the lighter pentomic and heavier MOMAR divisions. The new concept was presented to the Department of the Army in March 1961. The new army chief of staff Gen. George H. Decker approved it in April, and President Kennedy did so in a special message to Congress on May 25, 1961, describing it as essential to his flexible response strategy. In January 1962, the secretary of defense ordered the army to shift to the renamed Reorganization Objectives Army Division (ROAD) 1965. Although the Berlin Crisis delayed implementation until fiscal year 1964, ROAD represented a return to the triangular division of World War II, with a brigade headquarters rather than a regimental one. The brigade headquarters was designed to command two to five maneuver battalions (various types of infantry and armor) that could be attached or detached in a building-block concept. It was designed to be flexible so the army could organize forces as needed for specific missions, yet the army would maintain separate mechanized, infantry, airborne, and armored structures. Even the battalions were designed this way, allowing for the attachment or detachment of individual companies.[55]

The adoption of the ROAD division was not without its critics. Some complained that it required too much equipment, and others, including Taylor, thought it too soon to change from his beloved pentomic division. Later, tactical leaders further adjusted their units in the Vietnam War to better reflect reality. Likewise, during the early 1960s, the army designated heavier armored and mechanized formations for European deterrence, reserving light infantry and airmobile formations for limited counterguerrilla conflict, all but abandoning the modular building-block concept. Nevertheless, the army changed its basic organization for the second time in five years, marking the end of the short-lived pentomic division.[56]

The reorientation of the army for atomic warfare failed. The doctrine was never officially written, the equipment did not match the needs of the pentomic division, and the organization was cumbersome despite its smaller numbers. Atomic reorientation was successful in keeping the army relevant throughout the period, yet Taylor and Gavin's insistence on nuclear capabilities overshadowed more pressing concerns. Joining the nuclear club helped resurrect the service's image by portraying it as a forward-thinking "modern" atomic force. But atomic weapons were also complex, and without the proper mobility and communications equipment to put it all together, the pentomic experiment failed. The army's orientation toward atomic warfare reflected a broader trend in American military history in which the services placed an unbounded faith in technology to try to eliminate the human costs of war and make up for personnel shortcomings against much larger foes. Whereas airpower theory sought to eliminate ground combat, the army still accepted the need for ground soldiers yet also placed undue faith in technology—atomic weapons—to make up for personnel shortcomings. Critics also accused the army of fixating on expensive projects with little utility rather than updating the equipment of the average soldier.[57]

As President Kennedy took office in 1961, his administration implemented changes necessary to meet his conceptualizations of flexible response. Nevertheless, as soon as 1961, Secretary of Defense McNamara put into action a plan to increase the size of the regular army and refocus training for conventional ground warfare and counterinsurgency operations while letting the air force and NASA assume control of space and missile technology. The 1962 edition of FM 100-5 reflected that reality, reminding officers that "military forces must be able to operate effectively across the entire spectrum of war"—much like Kennedy's strategic vision. The ROAD division also reflected these ideas as it sought to provide increased flexibility with its seemingly infinite number of plug-and-play combinations, like the airborne division headquarters of World War II. The concomitant emphasis on Special Forces, strategic response elements, and light heliborne airmobile units also reflected the likelihood of this type of warfare. Although the army retained its nuclear-capable weapon systems, "the time and resources devoted to training to fight on the nuclear battlefields" were reoriented toward conventional combat and guerrilla warfare. A renewed focus on the equipment required to carry out mechanized warfare in Europe meant that the influence of airborne officers on the operational army was waning. The ROAD division also reflected the design structure of World War II–era armored divisions and their "combat command" brigade-like structure, a potential sign of the waning influence of airborne officers in the operational army.[58]

The pentomic period reflected both the influence of the airborne mafia and their shared values, beliefs, and norms about the efficacy of flexibility, dispersion, and mobility. "Each of the armed services has its own particular military philosophy . . . [about] how wars should be fought," as Secretary of Defense Wilson testified to the Senate Armed Services Committee in 1956. The army's philosophy was based on its leaders' experiences with airborne units in World War II. This shared experience played a monumental role in carrying out robust institutional changes in the atomic era, many of which proved ill-advised. "It takes courage to make them [innovations]," Gavin wrote, "because, for the few who will support a new concept, there will be hundreds who will point out why it cannot possibly work." In Gavin and Taylor's view, further advances in airlift capabilities, both tactical and strategic, promised to provide the service with the mobility required for the atomic battlefield while reinforcing key principles of airborne culture.[59]

CHAPTER 5

Tactical Mobility and the Airmobile Division

> The helicopter is aerodynamically unsound. It is like lifting oneself by one's bootstraps. It is no good as an air vehicle and I am not going to procure any. No matter what the Army says, I know that it does not need any.
>
> —Unnamed air force general, 1950

At 7:27 a.m. on February 24, 1991, the 101st Airborne Division's 1st Brigade flew ninety-five miles into Iraq to seize a swath of desert soon called Forward Operating Base Cobra. Following on the heels of their attack aircraft, the Screaming Eagles swiftly established a secure "nest" from which the division could deploy aircraft and infantrymen to screen opposing Iraqi forces along Highway 8 during the one-hundred-hour Operation Desert Storm. Cargo helicopters brought artillery, antitank assets, and fuel around the clock. Some two hundred thousand gallons of airlifted fuel allowed AH-1 and AH-64 attack helicopters to maintain a continuous armed presence along the highway. The next day, the division's 3rd Brigade launched more helicopters, which placed five hundred infantrymen along the route, effectively cutting off support between Baghdad and Kuwait. The operation doomed Iraqi forces in less than thirty-one hours. Like American cavalry of the mid-nineteenth century, the 101st Airborne Division executed a bold maneuver to seize key terrain, cut lines of communication, screen against reinforcements, and (with the French 6th Light Armored Division) protect the coalition's left flank. Not only did infantrymen deploy as dragoons, but attack helicopters operated as highly mobile mounted cavalry. The boldest air assault the US Army had ever attempted was a mission that the

101st, according to its commander, Maj. Gen. J. H. Binford Peay, "had trained for, for years."[1]

Army airmobility is, in essence, two interrelated concepts: (1) entirely aerial maneuver forces conducting screening, ambush, reconnaissance, and other traditional mounted cavalry missions in advance of ground units, and (2) "dragoons," or infantry maneuvering by air and fighting dismounted. These ideas developed from airborne warfare in World War II and the subsequent domination of army thinking by an air-minded group of officers. To the airborne mafia, the helicopter could provide the tactical mobility that paratroopers lacked in airborne operations throughout the Second World War and make up for the further lack of mobility exhibited in the Korean War. Maj. Gen. Harry W. O. Kinnard, commander of the army's first airmobile division from testing through to combat in Vietnam, said years later, "I just don't think I could have gotten the same kind of response, same kind of motivation, the same kind of understanding of flexibility of mind, of fighting in the enemy's rear" from non-airborne personnel.[2]

Airmobile, helicopter-borne infantry was the natural evolution of parachute infantry, just as some officers came to view parachute forces as the next evolution of the horse cavalry. This notion is due to the "mobility differential" provided first by horses and then by fixed-wing aircraft delivering parachute and glider forces to the battlefield. Horses provided a vast differential in speed and mobility over foot-borne infantry from the beginning of recorded history until the advent of the internal combustion engine. Until then, horses were the only way to move faster, further, and deliver more shock. By the Second World War, every motorized unit moved at the same speed, and the cavalry had lost its mobility differential. By late June 1950, horse cavalry was officially abolished and folded into the armor branch.[3]

In World War II, airborne forces were the only units with a higher degree of mobility than mechanized and motorized ground forces. Accordingly, airborne missions often took on a decidedly cavalry feel. During Operation Overlord, airborne forces secured the flanks of the Allied landings. Market Garden was a classic deep penetration operation to seize key terrain. Husky was a critical screening operation. Parachute and glider units provided commanders excellent operational reach but were relatively immobile and had to rely on attached transportation assets once on the ground. Operations in World War II demonstrated the shortcomings of parachute units yet promised the potential for greater mobility with improved technology. Airmobile warfare promised to address most of those limitations and give commanders more capability for vertical envelopment. Furthermore, the demands of theoretical tactical nuclear combat meant that battlefield mobility was vital to survival, and one way to quickly disperse and reassemble the battle groups of the pentomic division was with helicopters. Likewise, as the American commitment to Vietnam increased, so did the idea of using helicopters in that environment. Much as it had with airborne forces in World War II, the army sought a highly mobile quick-strike force to support armored and mechanized advances and which was versatile enough for a multitude of missions. The air cavalry that emerged during the Cold War is thus a derivative, a merger, of two schools of thought: the horse cavalry of old and the airborne forces of World War II. Airmobility doctrine resulted from airborne officers' wartime experiences and commitment to harnessing vertical maneuver; therefore it is a direct cultural descendant of parachute warfare in World War II.[4]

Sky Cavalry

The Korean War, according to General Gavin, highlighted the US Army's lack of quick-hitting cavalry forces to skirmish, patrol, screen, raid, and conduct reconnaissance for the US Eighth Army. Gavin published an influential article in the April 1954 issue of *Harper's* that he sent directly to the publisher to avoid Department of Defense censorship. He argued that the introduction of armor and subsequent mechanization of cavalry units after World War II had subverted the distinctive historical role of cavalry—it no longer enjoyed the essential advantage of speed and mobility over other forces. "With the motorization of the land forces and the consequent removal of the mobility differential," he wrote, "the cavalry has ceased to exist in our Army except in name." Through his *Harper's* article and other writings, Gavin called for a revolution in tactical mobility to be brought about by his "sky cavalry" concept, which he wanted "matched with highly mobile nuclear missiles . . . to operate over many thousands of square miles and be capable of gaining tactical decisions in a few hours." To Gavin, it was the perfect marriage of firepower and maneuver. In his estimation, sky cavalry could have been decisive in Korea.[5]

Soon after World War II, Gavin realized the obsolescence of the airborne division. "The airborne division is an outgrowth of the combat need for a hypermobile force in World War II, and the available commercial air transport to provide that mobility." The advent of improved surface-to-air missiles had rendered mass parachute drops obsolete. Gavin instead advocated the conversion of airborne divisions to sky cavalry units that maintained nominal parachute capability for greater flexibility of mission should a parachute assault be required. In addition, he believed all non–sky cavalry divisions should be armored ones with a preponderance of tanks and armored personnel carriers. Gavin tested the value of rotary-wing, vertical takeoff and landing (VTOL) aircraft firsthand as his 82nd Airborne Division was selected to test and evaluate thirteen Bell helicopters in 1946. While he first thought they might replace artillery-spotting aircraft, he soon learned that "[we] had in our hands an air vehicle of great versatility" and urged procuring more, including improved types.[6]

Some key leaders shared Gavin's views about the helicopter's potential. In a 1952 interview with the *New York Times*, the chief of staff of the army, Gen. J. Lawton Collins, explained that the army was considering helicopter and light planes to move soldiers and weapons "within a combat zone," intending to use the technology "to supplement and speed-up conventional truck-type transport and to make it possible to take men and weapons into otherwise inaccessible places on the battle line." The under secretary of the army, Earl Johnson, gave a rousing speech to the National Convention of the Air Force Association that surely ruffled some feathers. "The key to our plans must be mobility," he declared before outlining the army's great need for mobility to negate the Soviet manpower advantage. More mobility would allow the army "to hit our enemy three times for each blow we receive." Later in November, he stressed that one of the army's significant needs was organic aviation for "expediting and improving ground combat" and not for duplicating air force functions.[7]

In a May 1952 speech, the secretary of the army, Frank Pace, noted the importance of helicopter airmobility while explicitly linking developments in that field to the airborne mafia, noting that the airborne soldier was a "strange hybrid between soldier and airman." He continued, "From our airborne ranks have sprung a new breed of forward-looking commanders, including such men as Generals Matt Ridgway, Maxwell Taylor, and Tony McAuliffe, who have learned to think in terms of air and airmobility." Pace extolled the virtues of the helicopter, its role in replacing truck transport, and the importance of air ambulances in saving lives, using Korea as his case study. The helicopter's tactical mobility was also assumed to be critical to keeping units alive on the atomic battlefield. If atomic war ever materialized, it would require the dispersion of men and material in a manner only feasible through the advanced mobility brought by the helicopter. Heliborne infantry could also conduct ground reconnaissance, set up blocking positions, and harass columns while serving as a quick-reaction force to counter enemy forces until armored reserves could destroy them.[8] Airmobility was the wave of the future.

From 1954, Gavin used his position in the Pentagon as chief of army research and development to advance his theories on atomic warfare and, importantly, the helicopter. He published the *Harper's* article from this position and presented ideas for an entirely airmobile army. As Gavin wrote, the sky cavalry idea "came out of the Normandy experience" because he felt that airborne operations "had to mean something other than just parachutes and confusion." Airborne operations in World War II were an embryonic version of vertical envelopment that could evolve further with helicopters. Gavin saw his role in the June 1944 invasion in northwest Europe as screening in front of the main attack and tying up reinforcements—the role of the cavalry. Upon reflection, the paratroop general had realized that airborne operations had taken on the character of cavalry missions. In his postwar memoir, he assessed the similarities between the battle of Second Manassas and Normandy, finding them more common than he had realized in his studies. Yet he believed the airborne in World War II might have been even more effective with increased battlefield mobility. Processing his wartime experience, he foresaw that "in ground combat, the mobility differential we lack will be found in the air vehicle. Fully combined with the armored division, it would give us real mobility and momentum." To that end, Gavin urged the chief of Army Field Forces to relegate helicopter logistical efforts to secondary importance in favor of tactical heliborne operations, setting the stage for armed airmobile development.[9]

Airmobile ideas continued to spread around the army. Speaking at the 1954 Fort Benning Infantry Instructors Conference in June, Col. Joseph W. Stillwell Jr. declared that "the helicopter is a means of giving us the mobility that is so vital to atomic warfare. . . . [It] will allow us to disperse our units, offering unprofitable targets to enemy A-weapons and then allow us to mass our forces for a decisive blow at the critical time and space." A March 1955 *Collier's* article proclaimed the helicopter capable of making one American GI "equal to several Ivan's [*sic*] through the mobility of airborne cavalry"—yet another means of offsetting Soviet numerical advantages. In July of that year, Brig. Gen. Carl I. Hutton, the first commandant of the Army Aviation School at Fort

Rucker, Alabama, contemplated in *Army Aviation Digest* the various organizations of fighting aircraft that might help the army on the battlefield: "There might be, for example, a light, high-speed reconnaissance group, a fast striking force, an element to deliver a firepower punch, and finally a heavy fighting unit. The commander would coordinate the employment of the various fighting elements in the same way as an infantry or armored division commander." Throughout the mid-1950s, Gavin's ideas continued to grow at all levels of the army.[10]

One of Gavin's most important decisions while serving as the deputy chief of staff for operations (G-3) was his appointment of Brig. Gen. Hamilton Howze as the first director of army aviation. Selected because of his old-cavalryman's-style lifelong interest in mobility, Howze assumed the role in February 1955 and set out to make a strong case for procuring and organizing sky-cavalry-type units. Everywhere he went he made an aggressive case for more armed helicopters, telling the American Helicopter Society's 1956 annual forum that the "most spectacular and perhaps most important use of the helicopter is in attack and counterattack." Howze drew on scenarios borrowed from Command and General Staff College war games to make his case around the Pentagon. Instead of an armored division, he proposed a reinforced air cavalry brigade in Bavaria, defending against a hypothetical Soviet armored offensive. Without friendly tanks, the Americans destroyed bridges and roads and concentrated fire on chokepoints along likely avenues of approach. He assumed the same amount of air force fighter-bomber support as in the original war games. He added helicopter-mounted antiarmor weapons (still undeveloped) and helicopter-inserted infantry hunter/killer teams with antitank weapons to engage and destroy the adversary in-depth, with fewer losses than an armored division.[11]

Howze embodied the airborne, cavalry, and aviation cultural amalgamation from which the air assault concept emerged. He had spent the first ten years of his career with the old horse cavalry. He served with the 1st Armored Division in World War II, led a tank battalion and regiment, and attended an abbreviated flight school as a one-star general. After his tenure on Gavin's staff promoting aviation, he took command of the 82nd Airborne Division—his first assignment to a parachute outfit was as its commander. He later returned to Fort Bragg to take command of the XVIII Airborne Corps. By taking flight training as a senior combat arms officer, Howze helped the army impart the importance of aviation as he provided a general officer advocate for aviation officers. He was also part of the army elite—a West Pointer and protégé of Gavin and Chief of Staff Maxwell Taylor. Like many officers in the army after Korea, Howze believed that the air force had "flown away" from the battlefield and was enamored with high-altitude, supersonic aircraft that could deliver nuclear payloads rather than supporting ground soldiers. By 1960, some soldiers were convinced that no air force aircraft carried conventional bombs and that the Tactical Air Command had become Strategic Air Command in miniature. To Howze, they "had lost interest in the slow, low regime of flight—flight close to the treetops." The army still needed close air support. Howze realized that not all future combat would include atomic weapons or intercontinental ranges and that the army needed aerial platforms to deliver firepower and personnel from low altitudes, while also seeing the helicopter's potential in guerrilla warfare.[12]

Emphasis by senior leaders helped bring officers of other combat arms into aviation. John J. Tolson served with the 503rd Parachute Infantry in the Pacific and as an airborne planner on MacArthur's staff. By the mid-1950s, he was a Gavin acolyte in the Pentagon who later commanded the 1st Cavalry Division in Vietnam and the XVIII Airborne Corps. (And after retiring, he penned the seminal work on the history of airmobility.) When Tolson arrived at Gavin's G-3 office, he had just finished studying the future of the army aviation program, showing a deep understanding of the concept's history and the situation in 1953. He noted that aviation's purpose was to support land warfare through improved mobility, command, control, logistics, and greater battlefield dispersion for atomic war. Tolson advocated forming an "Army Airborne Corps" of personnel focused on army aircraft, rockets, and missiles over combat zones. This required "an entirely different connotation attached to the word 'Airborne' as to that used in the past." After Tolson's Pentagon stint, Gavin got him assigned to the Fort Benning Airborne Department, renamed the Airborne-Army Aviation Department. Benning became home to its own experimental Sky Cav unit under Tolson and Maj. William A. Howell, another 82nd Airborne officer turned aviator in 1946.[13]

Multiple exercises throughout the mid-1950s helped make a case for further helicopter development. Exercise Sage Brush was intended to test atomic-capable division structures and included sky cavalry for the first time. Organized around the 82nd Airborne Division's reconnaissance company, the ad hoc helicopter cavalry formation known as the "provisional reconnaissance troop" included three elements: a reconnaissance and surveillance section, a small, heavily armed, blocking section, and an antitank section. Heavy rains during the exercise left all forces road-bound and demonstrated the insufficiency of ground mobility in adverse weather. One observer, noting the lack of mobility, assessed that "a completely air transportable, more mobile division, air supplied to small units, and better communications would lead to an atomic capability." Tolson noted that the success of the sky cavalry during the exercise was due to the organic nature of its aviation. Like truck transportation units, regular army aviation assets were often sent to the units they were supporting, just before an operation, with the aviators unfamiliar with both the tactical situation and the assault unit. Developing units with dedicated aviation that trained consistently with the infantry and artillery units became a critical hallmark of the future airmobile division.[14]

The exercise also highlighted the air force's preference to deliver atomic weapons at the expense of the army's requirement for all-weather close-air support. Air force personnel in charge of the overall airspace delayed clearance for so long that army pilots often took off without authorization, and one senior army officer told his pilots to ignore all air force directives. Exercise director Maj. Gen. Paul D. Adams noted that "the requirement for close air support . . . continues to be recognized as of paramount importance and that efforts be directed toward the provision of aircraft, and air delivered ordnance, which will meet precisely the needs of Army forces in the immediate vicinity of front line units, as well as the need for extensive and quickly responsive air reconnaissance of the entire battle area." Increased emphasis on helicopters as an armed close-air support platform and troop carrier could solve these issues.[15]

As director of army aviation, Howze gave tacit approval to the efforts of Brig. Gen. Carl I. Hutton, commandant of the Aviation School at Fort Rucker, Alabama, to experiment with armed helicopters in 1956. CONARC commander Gen. Willard G. Wyman authorized these experiments in June. Hutton appointed Col. Jay Vanderpool to lead the effort, and he and his small team scrounged excess hardware and surplus personnel to form a "sky cav" platoon that tested various combinations of machine guns and rockets fired from helicopters. "Vanderpool's Fools" armed Bell OH-13 helicopters with .50-caliber Browning M2 machine guns and Swiss-made Oerlikon 8-centimeter antitank rockets and successfully fired both weapons on July 5, 1956. To do so, they "violated numerous regulations, begged, borrowed (and when necessary) stole material . . . to sell the concept of sky cavalry based on the armed helicopter." The tests were done in secrecy, without money, and on nights and weekends to not arouse suspicion from the Pentagon or air force personnel keen on ensuring the army did not usurp the 1948 Key West Agreement. Vanderpool and his team continued to test mounted helicopter weapons, from M60 machine guns to 20mm cannons and 2.75-inch rockets. Meanwhile, Gavin had Tolson assigned as director of the Airborne Department at Fort Benning, where he was given "instructions to develop tactical doctrine for the combat employment of helicopters." While Vanderpool tested armed helicopters, Tolson and Howell considered ideas for delivering infantrymen to the battlefield. The teams at each location frequently shared knowledge.[16]

Explicitly connecting helicopters to the old horse cavalry, Vanderpool wrote that their organizational theory reflected their understanding of how the Duke of Wellington employed cavalry. "His cavalrymen fought from their mounts," Vanderpool wrote. "The dragoons were horsemounted infantry who dismounted to fight. They were supported by horsemobile artillery. His trains moved by horsedrawn wagon." Everything then was horse-mounted, so every aspect of new sky cavalry units would also be airmobile. These ideas drove army thought on airmobile tactics so forcefully that the team at Fort Rucker rewrote the 1936 horse cavalry field manual using the word "helicopter" instead of "horse." Airmobility was meant to resemble horse cavalry of old so that "older soldiers, I mean two-, three- and four-star generals, could understand the language of their day, of the late [19]30s. It did help sell the concept." Finally, in October 1959, Gavin's successor as chief of research and development, Lt. Gen. Arthur G. Trudeau (onetime 1st Cavalry Division commander), ordered army aviation leaders to develop plans to bridge the gap between army and air force responsibilities and develop aircraft to meet army requirements.[17]

The Command and General Staff College defined the term "airmobility" in a June 1959 article in the service's premier professional journal, *Military Review*. While airmobile helicopter troops did not need the special training of parachute troops, airmobile operations were often much more complicated to plan than regular ground assaults. The need for definition was largely to differentiate the concept from large-scale *joint* airborne (parachute) operations. First, these airmobile operations were defined as "those airlifted combat operations conducted by and within the Army" or, in other words, not in conjunction with the air force. Further, airmobile operations were generally smaller and more tactically oriented than airborne operations. Air and ground elements enjoyed a symbiotic, highly coordinated relationship, often from the same organic maneuver formation that worked together on an intimate, day-to-day basis. And finally, all major portions of the operation were army units commanded by a single land force commander. After this article, army doctrine used "airmobile" exclusively, and the 1960 edition of Field Manual 57–35 was simply titled *Airmobile Operations*.[18]

The Rogers and Howze Boards

Efforts by Gavin, Howze, Hutton, Tolson, and Vanderpool in the 1950s helped establish the Army Aircraft Requirements Review Board in 1960, chaired by Lt. Gen. Gordon B. Rogers (and better known as the Rogers Board). When the Korean War began, the army had only 668 light planes and 57 helicopters. By the Rogers Board's convening, it had over 5,000 aging, limited aircraft. The board was tasked with developing requirements and procuring new helicopters. As deputy commanding general for CONARC, Rogers received an explicit charter to review industry proposals using his 1959 Army Aircraft Development Plan. Rogers's team and the civilian aviation industry developed proposals for aircraft fulfilling three primary functions: observation, surveillance, and transport. In January, forty-five corporations sent 119 new concepts, including autogiros, helicopters, STOL (short takeoff and landing) and VTOL aircraft. The board's most significant contribution was recommending the turbine-powered Bell XH-40 utility helicopter, later called the UH-1 Huey. The Huey and CH-47 Chinook proved instrumental in modernizing Army aviation, providing platforms for the "air fighting units" that Howze included in the final report addendum. The board also designated replacing aircraft every decade. Its findings prompted little action beyond procurement, and was often overshadowed by the later, more successful Howze Board.[19]

As the new secretary of defense in 1961, Robert S. McNamara took a keen interest in army aviation. Upon review of developments to that point, he expressed his displeasure about the state of development in army aviation by way of two April 19, 1962, memorandums to the secretary of the army, Elvis J. Stahr. Denouncing the lack of action since the Rogers Board, McNamara implored the army to begin taking a "bold new look at land warfare mobility . . . divorced from traditional viewpoints," one that might give the army a "quantum increase in effectiveness." McNamara's memo dictated the method and suggested specific officers to tackle the task, including Howze and Col. John Norton—Gavin's wartime operations officer at regimental and division levels. The second memorandum, issued the same day, outlined subjects to explore to free the ground soldier "from the restrictions imposed by the earth's surface" and provided a September 1 deadline for the full report, with Stahr's recommendations. McNamara's memorandum called for revolutionary and unorthodox ideas to take the army forward and to accomplish this study in a mere four months.[20]

The memorandum was drafted by army officers who were apostles of airmobility—not McNamara himself. While McNamara is often given outsize significance within most historical treatments of the era, the culture of airmobility had already taken root. It was now being implemented in concrete actions by true believers. One of those officers was Col. Robert R. Williams, a pilot who began his career flying light artillery-observation aircraft in World War II. Williams had served in the G-3 as chief of the aviation branch before Howze's appointment. By 1962 he was serving as deputy director of the Tactical Warfare Systems Office in the Office of the Secretary of Defense. He had access to the secretary to write this memorandum alongside Col. Edwin Powell. According to Williams, McNamara signed it "almost exactly" as he wrote it. Finally, army aviation enthusiasts had a defense secretary who supported their vision. The memo established the US Army Tactical Mobility Requirements Board—better known as the Howze Board—which got to work in 1962 at Fort Bragg, North Carolina.[21]

While the Rogers Board proposed the design and acquisition of new equipment, the Howze Board was tasked to ascertain how to *use* that equipment. Led by Howze, now the commander of the XVIII Airborne Corps, the board consisted of many of the airborne-turned-airmobile mafia, including Tolson, Norton, and Williams. All told, the board comprised thirteen general officers and five senior civilian officials. The board was a mix of armor officers, sons of horse cavalry officers, and paratroopers, representative of the mix of talent and ideas spurring airmobility during the period. The US Air Force ensured it also had a general officer observer on the board, though he was not to participate in internal discussions.[22]

Throughout the board's deliberations during the summer, Howze sought input from senior officers and retirees to ensure no stone was left unturned, asking for their ideas on development over the next decade. The officers he solicited included Gavin, Creighton Abrams, William Westmoreland, and Harry Kinnard. Gavin provided a thoughtful, if far-fetched, input, envisioning the use of manned missiles to deliver troops around the world. At the same time, Westmoreland believed air vehicles would phase out ground vehicles in much the same way the internal combustion engine phased out horse transportation. In total, Gavin's ideas represented his belief in the need for a "mobility differential" for cavalry forces. Maj. Gen. Andrew Boyle thought actions in Vietnam to that point had already proved the feasibility of airmobility, while Abrams noted the decisive potential of a mobility differential and that "in the area of guerrilla warfare and counterinsurgency, it seems to me that it is already quite well established that forces built largely around these air vehicles are the most effective." Because all the board members were mobility believers who had supported airmobile concepts for years, its findings were never in doubt. The board was determined to demonstrate how the army might increase airmobility.[23]

Howze's board met in an empty elementary school at Fort Bragg throughout the summer and immediately got to work on a program of war-game simulations, field exercises, and research. The board had three infantry battle groups, artillery, and engineers from the 82nd Airborne Division at its disposal, alongside 150 fixed- and rotary-wing army aircraft. C-130 Hercules and close-air support aircraft provided by the air force's Tactical Air Command added realism to scenarios while allowing the air force to showcase its tactical mobility capabilities. The board's members used intelligence reports on Sino-Soviet developments and capabilities, reports from the Combat Developments Command and officers in Southeast Asia, and reports from logisticians, manufacturers, armament engineers, and scientists forecasting new weapons and equipment that could be developed over the next decade. The combined effort was designed to come up with ideas tailored explicitly to reality.[24]

Howze had only ninety days to conduct hundreds of tests, from squad tactics to division and higher-level transportability and logistics. Howze and his hundred board members ensured that they tested concepts in various conditions against myriad potential adversaries in multiple geographic conditions. They studied the role of helicopter-mounted combat as an extension of traditional cavalry concepts in simulated operations in Southeast Asia, Northeast Asia, Europe, and the Middle East. Outside of war gaming, the board performed more than twenty live tests at Fort Bragg, across agricultural areas and hills of central and western North Carolina, in the swamps and soft sandy conditions of Fort Stewart, Georgia, and in the forests of western Virginia to simulate

jungle. Applications to counterinsurgency proved promising yet difficult. Despite the promising tests, they "could devise no sure tactic . . .
to ensnare an enemy guerrilla force." The final trial occurred in front
of the secretary of defense, the secretary of the army, and the army chief
of staff in an impressive show that helped convince these leaders of the
possibilities that airmobility promised.[25]

After the demonstration of airmobility in atomic, conventional, and
counterguerrilla situations, the final report simply concluded that "operational tasks can be done better with than without army aviation."
The board also stated that airmobile units would be ideal for "fighting
Communist armies in Southeast Asia or Korea" because of the units'
extreme mobility amid rugged terrain. Ultimately, Howze's final thirty-
five-hundred-page report, delivered on August 20, 1962, concluded that
"adoption by the army of the airmobile concept—however imperfectly
it may be described and justified in this report—is necessary and desirable" and likened the transition to "that from animal mobility to mo-
tor." The board also determined that airmobile assets would enhance
ground forces' combat effectiveness.[26]

Logistics was another avenue of significant testing. The board proposed using assigned army aircraft, both fixed- and rotary-wing, to
move supplies forward. The air force was critical to this operation and
envisioned itself as the "wholesaler" bringing supplies from the rear
forward to the limits of its capabilities, often via C-130s on unimproved runways. From there, the army was the "retailer," pushing supplies where needed: Caribou transport aircraft and Chinook helicopters
would bring them further to the brigade level, where teams would ear-
mark and distribute the supplies to even lower echelons. Hueys would
transport those teams and equipment down to the battalions and companies where needed. Every logistics element was airmobile; even maintenance and supply sections had H-37 Mojave helicopters—the largest
rotary-wing aircraft in army inventory.[27]

The Howze Board proposed five new organizations, but only the air
assault division was implemented. It boasted 459 aircraft, a significant
increase from the fewer than 100 of a standard infantry division. To reduce its strategic airlift requirement, the new airmobile division would
have only 1,100 ground vehicles—down from 3,452 in a standard infantry division. The air assault division had significantly reduced artillery,
relying on just three 105mm howitzer battalions and Little John rockets, all air transportable in Chinook helicopters. However, these were
augmented by twenty-four OV-1 Mohawks and thirty-six armed Huey

helicopters. The Mohawks—fixed-wing turboprop observation and re-connaissance aircraft—were later sacrificed to placate the air force's concerns over roles and missions. Despite the increased number of lift helicopters, the division's organic assets could only move a brigade at a time, whereas the proposed but never implemented air cavalry combat brigade was to have a seat in a helicopter for every member of the unit. Regardless, the air assault division's streamlined organization reflected the similarly sleek structure of early airborne divisions. This likewise reflected the belief among airborne officers that their men were innova-tive and adaptable enough to accomplish their mission with minimal support.[28]

The air cavalry combat brigade best reflected Gavin's sky cavalry idea. It was to have 316 helicopters and function like traditional cav-alry. Half of its aircraft would be gunships. This was a force to screen, reconnoiter, and wage delaying actions, all from the air. They were to fight like the cavalryman on horseback, from an aerial mounted posi-tion, and perform the historical role of cavalry in exploitation, pursuit, counterattack, and flank protection. The board also proposed a corps aviation brigade (207 aircraft) to allow rapid movement of reserves and equipment; an air transport brigade (134 aircraft) to support the air assault division logistically, and a special warfare aviation brigade (125 aircraft) to support Special Forces units in combat. In addition, the board recommended increases in organic aviation to all division types, a lofty and expensive goal. Howze's board also determined that the cost of an airmobile division was $186 million more than an armored divi-sion over five years and $294 million more than a standard infantry division. Despite those numbers, Howze recommended that the army adopt five total air assault divisions, three air cavalry combat brigades, and five air transport brigades. This proposal to support five air assault divisions within a sixteen-division force "curled a lot of hair among staff officers in the Pentagon," according to Howze.[29]

In his fiscal year 1964 budget proposal, published in December 1962, Secretary McNamara decided not to pursue the entirety of the recom-mendations put forth by the Howze Board. He wrote to the president that "full implementation of the Secretary of the Army proposals at this time would be premature" and that the Howze Board "did not take full account of how the Air Force might contribute to Army tactical mobil-ity." One of the foremost reasons for caution was fiscal constraints. If the army adopted every portion of Howze's program, it would procure six thousand aircraft between 1963 and 1968, costing $3.8 billion. In the 1964 fiscal year, that amounted to $1.12 billion instead of the previous $371 million earmarked for army aviation. McNamara decided to test just one division but instructed the army to draw up a five-year aviation procurement plan to prepare for any eventuality.[30]

Howze and the commanders of the 82nd and 101st Airborne Divisions argued for the immediate conversion of those divisions to airmobile ones that retained a parachute capability. McNamara, however, did not want to degrade the readiness of the nation's core quick-response divisions within the strategic reserve. Instead, he proposed testing one division and developing a robust five-year aviation plan for how the army would continue to take advantage of helicopter technology. Howze was called away in October during the height of the Cuban Missile Crisis to serve as the potential Cuban invasion force commander. Rather than adopt the division at that moment, army leaders decided to continue testing airmobile concepts. The army chief of staff Gen. Earl G. Wheeler and the new secretary of the army Cyrus Vance echoed McNamara's sentiments and ordered the creation of a test division. Of course, with the looming shadow of Vietnam in mind, the helicopter seemed well suited for moving infantry around the difficult terrain found in Southeast Asia.[31]

Test Division

Despite Howze's best efforts, many on the army staff were still not convinced that activating a full-fledged, expensive air assault division was worth the cost. The chief of staff, Gen. Earle Wheeler, wanted to see more tests—of different unit sizes, from company up to division. To do so, the deputy chief of staff for operations issued guidance on January 7, 1963, for the organization, training, and testing of one air assault division with an attached air transport brigade. On January 18, the army announced a three-phase testing program to take the concept through various trials, from platoon to division level. The 11th Air Assault Division was then activated at Fort Benning on February 15, around a small cadre of officers and men at the Harmony Church area of the base. According to Howze, the 11th was "to develop the details of doctrine, tactics, and technique for its [the airmobility concept's] employment." Choosing the 11th Airborne Division as the test unit indicated the impact of the airborne mafia on the development of airmobility and a shared concept of "air-mindedness" predicated on light, easily transportable units and equipment. The division was charged with further testing concepts developed the previous summer at Fort Bragg, but now with a dedicated unit, personnel, equipment, and a direct cultural link to airborne warfare.[32]

The army also activated the 10th Air Transport Brigade to provide thirty-two additional CV-2 Caribous, nineteen CH-47 Chinooks, and three CH-54 Skycranes to move the division, keep it supplied, and evacuate casualties. The army reassigned Brig. Gen. Robert R. Williams from his position with the chief of staff to oversee and assess everything as the test-and-evaluation group leader at Fort Benning. Williams reported directly to Lt. Gen. Charles W. G. Rich, who, as the commander of Fort Benning, became the overall test director for Project TEAM (Test and Evaluation of Air Mobility). Thanks to the appointment of Brig. Gen. Jack Norton as assistant commandant of the Infantry School, the division met little bureaucratic resistance.[33]

Wheeler selected Brig. Gen. Harry W. O. Kinnard to command the new division. During World War II, Kinnard served as a battalion commander in the 101st Airborne Division and later as the division operations officer at Bastogne. Kinnard also played an instrumental role in securing the legacy of airborne units when he helped produce the 1949 Hollywood hit *Battleground!* Every operational assignment Kinnard held was in the airborne. He was the 101st's assistant division commander when summoned to take over the 11th Air Assault Division (Test) at Fort Benning. The élan, can-do attitude, flexibility, and expeditionary mindset of World War II airborne units undergirded the development of airmobility thanks to the selection of Kinnard as the first commander of the test division. His division started as a shell of its future self—just a single infantry battalion—before adding more units and absorbing five battalions (three infantry, two artillery) of the 2nd Infantry Division stationed at Fort Benning for testing. The pace was quick—not unlike airborne testing in the early 1940s. By June 1964, the 11th Air Assault Division (Test) had practiced battalion, brigade, and division-level maneuvers.[34]

The airborne mindset played an enormous role in how Kinnard led his division. He was so convinced of the importance of airborne officers that all battalion and brigade commanders in his division had to be airborne qualified. Kinnard firmly believed that "airmobility is a state of mind and I found that, by and large, parachutists are more able to adapt to that state of mind than other people." According to some of his officers, he spent much of his time talking about this "state of mind" and tried to instill an ethos of "air-mindedness" to make every aspect of the unit airmobile. This manifested in leaders exiting their helicopters first onto the landing zone, just as paratrooper officers exited their aircraft in front of their men. This was an example of a unit culture that insisted leaders share the same hardship as their men, and often more. He preached the importance of "squeezing the excess weight out of the division," much like early airborne units, and reiterated to his men that the air assault was the next step for the army to take. Just as the World War II–era airborne division needed to be streamlined to maximize air transportability, so did the pre-Vietnam air assault division.[35]

As the division commander, Kinnard was given the freedom to hand-pick his subordinates and, like the airborne innovators twenty years before, gave them a wide latitude to accomplish their goals. To further implant his airborne mentality on the division, Kinnard ensured that parachute-qualified officers commanded subordinate units, and many enlisted paratroopers found their way to his division. The division's 1st Brigade was officially designated as airborne and formed around the 1st Battalion, 187th Infantry Regiment—one of the 11th's most storied units and the only regiment to make parachute jumps in Korea. One of the officers Kinnard picked was Lt. Col. Harold G. "Hal" Moore. Having tested experimental parachutes under Kinnard's command, Moore wrote Kinnard asking to take command of a battalion. In April 1964, the Pentagon ordered the younger officer to report to Fort Benning immediately to take command of the 2nd Battalion, 23rd Infantry. Moore had served as the "one-man airborne branch in the Air Mobility Division for two and a half years" at the Pentagon Office of the Chief of Research and Development. There he worked for Gavin, Howze, Norton, and other airmobility disciples, giving birth to the concept.[36]

In another instance, Kinnard removed Lt. Col. A. J. Millard, a highly competent, well-loved commander in charge of the 2nd Battalion, 38th Infantry, merely because he lacked jump wings. His replacement, Lt. Col. Robert Tully, was a veteran of the two combat jumps in Korea. Another paratrooper, 3rd Brigade commander Col. Thomas W. Brown, had graduated from West Point in January 1943 and served in World War II with the 11th and 13th Airborne Divisions. Like Kinnard, Brown was known for giving his subordinates "the freedom to run their units," displaying comfort in the sort of decentralized operations he had grown accustomed to in the war. He would need to be even more comfortable commanding with his airmobile brigade.[37]

To engender esprit de corps and an airmobile state of mind, Kinnard authorized an air assault badge. The badge was designed to duplicate

paratrooper and aviator wings and was earned by successfully rappel-
ling from a helicopter three times at 60 feet and twice at 120 feet. Many
longtime paratroopers thought "it was a whole lot scarier hanging from
a rope from a hovering helicopter than it ever was just jumping out of
an airplane." However, the badge never earned Department of the Army
approval and was terminated when the 11th Air Assault became the
1st Cavalry Division. However, during its short life, it became a coveted
uniform item. A different air assault badge finally gained approval in
1978. Today, the badge is worn with the same background trimming as
paratroopers wore in World War II.[38]

Kinnard's efforts to instill unit pride were one thing, but his division
still had to pass muster in various exercises. Initial large-scale testing
began in September 1963 with Exercise Air Assault I, during which
the reinforced 1st Battalion, 187th Infantry Regiment, conducted
battalion-size air assault maneuvers at Fort Stewart, Georgia. During
these initial tests, Kinnard and his division realized they encountered
multiple problems, including inadequate signal equipment, insufficient tables of organization, and helicopter defects. The new Boeing-
Vertol CH-47 Chinook had quality-control issues during production,
making the airframe unreliable and spare parts scarce. Throughout
1963 and 1964, division maintenance personnel worked directly with
Boeing to improve the aircraft's reliability and turn it into one of the
army's most versatile and reliable helicopters. The CH-47 was critical
because of its lift capability—carrying up to forty-eight troopers or ten
thousand pounds of cargo. Kinnard also had to bend the rules and
sometimes outright ignore regulations, as flying in formation or close
to the trees was unauthorized at the time. Nevertheless, the 11th Air
Assault used the period to push the limits of how far the army could
go with helicopters. In November 1963, the secretary of the army Cyrus
Vance referred to the efforts of the 11th Air Assault as the most important in the army.[39]

Critics of the helicopter cited the aircraft's vulnerability and inability to fly at night or in adverse weather. Gavin always thought
the helicopter's vulnerability meant it should avoid heavily defended
areas. He used to argue with John Norton and Harry Kinnard in the
Pentagon—Gavin "felt it was cavalry, and it shouldn't be used where
it's extremely vulnerable; it should be used to give the commander information and time and a little space." However, Kinnard and Norton
envisioned a far more capable, entirely air fighting unit. These ideas

were tested during a massive validation exercise in the fall of 1964 dubbed Exercise Air Assault II. Hurricane Isbell pounded the Eastern Seaboard with thunderstorms on the morning of the exercise, and low ceilings cloaked the region. The air force grounded its aircraft alongside commercial flights on the entire coast. Despite the conditions, 120 helicopters of the 11th Air Assault launched only one hour behind schedule, placed a battalion of troopers on their objectives, and validated the inclement weather capability of army helicopters. Air Assault II lasted almost two months, involved thirty-five thousand troops, and took place across four million acres of the Carolinas—the largest peacetime exercise since the similarly located Carolina Maneuvers in 1941.[40]

To the officers observing the exercise, Air Assault II demonstrated the efficacy of the entire airmobility concept. The new army chief of staff, Gen. Harold K. Johnson, visited the exercise on November 4–5. After seeing the competing Gold Fire I air force exercise as well, he commented, "I had the rare privilege of seeing the 11th Air Assault one week and the other concept at the early part of the following week, and I would make a comparison of perhaps a gazelle and an elephant. The two are not comparable." Johnson was referring to the inability of the air force exercise to demonstrate the tactical mobility offered by swarms of UH-1 Hueys across the Carolinas.[41]

Another critical factor in developing airmobile doctrine was the real-world experience of army aviation in Southeast Asia. Airmobility seemed ideal for offsetting the advantage Vietnamese insurgent forces enjoyed in surprise and intelligence. Kinnard and other army leaders recognized this, remembering that the "Viet Minh had defeated a well-trained, well-equipped, *ground-bound* French force in the early 1950s." To them, airmobility promised to be the decisive difference in the American effort. Multiple transport helicopter companies were already committed to Vietnam, providing direct support to South Vietnamese units and their American advisers. These pilots pioneered concepts in formation flying, air assault fire support, and logistics in a real-world combat scenario and disseminated that knowledge to the 11th Air Assault Division in the United States. Gen. Earle G. Wheeler testified to Congress in 1963 that "the greatest benefits [of airmobility] accrue in the counterinsurgency and limited war situations" and in 1964 wrote that "the addition of the helicopter gives the Army of the Republic of Viet Nam a mobility differential and the advantage that goes with it."

The army was enamored with the potential of helicopters to fight a counterinsurgency campaign.[42]

Many pilots with combat experience in Southeast Asia returned to join the 11th Air Assault Division. Likewise, officers and aviators from the 11th visited units in Vietnam to continue the cross-fertilization of ideas and information. Lt. Col. Moore remembered years later that by May 1965, he and the rest of the division's battalion and brigade commanders, along with their staffs, received after action reports of the 173rd Airborne Brigade's engagements and "were reporting to heavily guarded classrooms at the Infantry School in Fort Benning, Georgia, for top-secret map exercises. The maps the games were played on covered the Central Highlands of South Vietnam." Division leaders quickly realized that all signs pointed to the employment of their airmobile division in Southeast Asia.[43]

In March 1965, the army converted the understrength 11th Air Assault Division (Test) to a full-strength active army division within the existing force structure. But it was not until June 16 that McNamara publicly announced its inclusion as part of the sixteen-division army. The decision was made for several reasons but mainly because the army needed an airmobile capability to meet various threats, and the test division had proven, through training scenarios, that it could be used effectively in anything from general nuclear war to low-intensity counterinsurgency. The situation in Vietnam played a critical role in its development. To some, the airmobile concept appeared ideal for maneuvering counterinsurgent forces around the rugged jungle terrain of Southeast Asia. The only problem was that counterinsurgency doctrine required securing and controlling the population on the ground, a difficult task for an organization meant to fly everywhere. Hutton realized this as early as 1963, and after studying the eventual failure of French tactics with helicopters in Algeria, he worried that the same might happen to the US Army and South Vietnamese forces. "Because the helicopter is inherently so effective," Hutton wrote, "these tactics do result in local successes. They fail strategically, however, because they leave the initiative to the guerrillas. They permit the guerrillas to choose the time and place of fighting, and they allow entrapment situations." Further, once the novelty wore off, antiaircraft gunners in Algeria and later Vietnam became adept at shooting them down. Nevertheless, Hutton thought they were still valuable—especially for logistics and sustained operations—but should be used with caution.[44]

Back to the Cavalry

After McNamara decided to keep one air assault division on active duty, Kinnard advocated for his beloved 101st Airborne to be named the army's first of its kind. He also desired that all combat arms personnel within the division be parachute-qualified—as Howze had proposed in 1962. Kinnard and Howze's belief in the air-mindedness of paratroopers and the cultural link to parachute units of World War II drove these ideas. However, that was not to be, and on July 1, 1965, the division reflagged as the 1st Cavalry Division (Airmobile), absorbing elements of the 2nd Infantry Division at Fort Benning while swapping division colors. The 2nd Infantry Division's standard went to Korea, while the 1st Cavalry colors moved to Fort Benning. The division was to have 15,787 personnel, 434 aircraft, and 1,600 vehicles, compared to a standard infantry ROAD (Reorganization Objectives Army Division) of 15,900 personnel, 101 aircraft, and 3,200 ground vehicles.[45]

The army's choice of the 1st Cavalry was by design. In 1965, that division was organized as a standard ROAD infantry division on garrison duty in South Korea. This fate was unsettling to old cavalrymen. By this point, many old horse soldiers from the 1930s had ascended to power in the Pentagon, superseding the airborne mafia; they included the army chief of staff, Gen. Harold K. Johnson. In Korea, Johnson had commanded one battalion and two regiments in the 1st Cavalry Division. Horses were gone, of course, and the cavalry branch was now armor. Still, these officers, who had been generally hostile to all forms of airmobility, now saw the helicopter as a return to mobility for the cavalry.[46]

When McNamara announced that the 1st Cavalry Division would assume the mantle as the army's first active airmobile division, he gave it eight weeks to reach combat readiness. Kinnard maintained command, and his division was authorized eight infantry and three artillery battalions, a helicopter aerial rocket battalion, an air cavalry squadron, and two assault helicopter battalions. However, when the order came to deploy to Vietnam, Kinnard only had 9,849 men and had to replace those ineligible for deployment because President Johnson did not declare a national emergency. Without emergency authorization, anyone who had recently returned from overseas duty or whose term of service was ending could not deploy. Meanwhile, the division made several tweaks to its organization, including eliminating fixed-wing Mohawk attack aircraft—a sacrifice to preserve air force relations—and the Little John atomic rocket battalion. Six Mohawks remained in the division for reconnaissance purposes but no longer as close-air-support attack aircraft. Later, in 1966, the army relinquished control of all CV-2 Caribous to ensure the air force ceased encroachment on rotary-wing development.[47]

The 1st Brigade, 1st Cavalry Division, remained on airborne status at the direction of the vice chief of staff, Gen. Creighton Abrams. This was Kinnard's primary brigade, the one that had initiated testing and carried the airborne lineage of the 11th Airborne with battalions from its original 187th, 188th, and 511th Infantry Regiments. These were now 1st and 2nd Battalions (Airborne), 8th Cavalry, and 1st Battalion (Airborne), 12th Cavalry. All things being equal, Kinnard would commit this unit to battle first. An entire brigade on jump status, however, required airborne-qualified personnel, so the brigade brought in experienced noncommissioned officers from the 101st and 82nd Airborne Divisions and set up a special iteration of the Basic Airborne Course to qualify another 659 paratroopers to round out the unit.[48]

The division was declared combat-ready on July 1, 1965. The 1st Cavalry Division was ordered to Vietnam, publicly announced on television by President Lyndon B. Johnson on July 28, 1965. The airmobile division seemed ideal for combat in the jungle terrain of South Vietnam, as if they were created explicitly for that purpose. As Creighton Abrams remarked, "Is it not fortuitous that we happen to have this organization in existence at this point in time?" The unit was scheduled to arrive in South Vietnam between September 9 and 18, 1965. The airmobile division began movement shortly thereafter, requiring a staggering amount of surface vessels to transport its equipment—six passenger vessels, eleven cargo ships, and four aircraft carriers. Its first elements arrived at An Khe on September 14, and troopers of the 1st Cavalry Division assumed responsibility for the defense of that area on September 28, 1965, three months after the unit was activated on July 1.[49]

Kinnard wanted to base his division in Thailand to operate up and down the Ho Chi Minh Trail in Laos and Cambodia. He wanted to use his division's ability to interdict quickly through deep penetration raids, to cut the logistical lifeline of the insurgency. Gen. William C. Westmoreland, commander of Military Assistance Command, Vietnam (MACV), denied this request, as it was outside the MACV area of operations. He also wanted to split the division's assets across different parts of the country, to flood the countryside with more men to buttress MACV pacification efforts. The two compromised, and Westmoreland assigned Kinnard's division the responsibility of securing the main line of communication, a road network, through the Central Highlands in Pleiku Province to prevent North Vietnamese forces from severing South Vietnam in two. No other unit in the army had similar capabilities. Airborne units had superb strategic mobility but were limited to whatever could be dropped by parachute in the battle area. The French had proved the limitations of such units at Dien Bien Phu in 1954. Standard infantry divisions in the mid-1960s relied on either mechanized or motorized transportation, and armored divisions were far too cumbersome for employment in a jungle environment with a paltry road network. The airmobile division seemed the perfect choice to fight in the challenging environment of Vietnam.[50]

The first full test of the airmobile concept came during Operation Silver Bayonet. This was the pursuit phase after a defense of the Plei Me Special Forces Camp in October from an assault by forces of the People's Army of Vietnam (PAVN). Elements of the 1st Brigade had rescued South Vietnamese forces at Plei Me and secured over twenty-five hundred kilometers of territory in the first battle of the Ia Drang campaign. Kinnard believed the action validated the nearly three years of hard work poured into testing and training his division. After the successful defense of the camp between October 19 and October 28, Kinnard received authorization to pursue Gen. Chu Huy Man's withdrawing elements. Kinnard's air cavalry squadron, commanded by Col. Richard Stockton, pursued the enemy, conducted a nighttime ambush that included helicopters firing within fifty meters of friendly ground troops, and turned the battle over to the infantry battalions. The action of Stockton's squadron closely resembled Gavin's initial "sky cavalry" concept and further validated the previous decade of conceptualization and testing. Between October 28 and November 14, 1st Cavalry Division troopers and Huey helicopters swarmed around Pleiku Province, searching for enemy forces. Kinnard wanted to keep pushing his elements toward the Cambodian border, and intelligence gathered suggested a North Vietnamese presence in the Ia Drang Valley near the Chu Pong Mountains.[51]

On November 14, 1965, the first of Lt. Col. Hal Moore's 1st Battalion, 7th Cavalry, landed in sixteen helicopters at Landing Zone (LZ) X-Ray. They arrived to begin the pivotal portion of the campaign after traditional artillery, aerial artillery (helicopter-mounted rockets), and helicopter gunships fired on possible enemy positions near the landing zone. The same sixteen helicopters would continue to move troopers onto the landing zone as Moore's men fought a gallant three-day battle that included the first-ever tactical use of the B-52 bomber. Their mission was to seek and destroy enemy main force units in the valley. Despite no activity having been detected in the immediate area, three PAVN battalions were headquartered on a ridge overlooking the landing zone. Three hours after the landing, PAVN forces attacked from their battalion redoubt on the mountain above the landing zone. This unleashed a brutal firefight that temporarily closed the landing zone just after Moore received reinforcements from Company B from the 2nd Battalion, 7th Cavalry. Moore's situation at one point became so tenuous that he issued the SOS call "Broken Arrow," meaning that every air force platform within reach should come to the aid of his unit, which was in danger of being overrun.[52]

The PAVN had stymied the airmobile division's key capability. With reinforcements by air cut off, reinforcements from the 2nd Battalion, 5th Cavalry, had to move in on foot from LZ Victor, three kilometers southeast of X-Ray. The fighting subsided enough on the morning of November 16 to allow the 3rd Brigade commander, Col. Thomas W. Brown, to relieve Moore's beleaguered men with the remainder of 2nd Battalion, 7th Cavalry—again on foot. The rest of the cavalrymen moved out the next morning toward two alternate landing zones—Columbus and Albany.[53] On the way to Landing Zone Albany, Lt. Col. Robert A. McDade's 2–7 Cavalry was ambushed by a PAVN battalion. The ambush achieved total surprise, and pilots attempting to provide resupply and support to McDade's troopers had difficulty distinguishing between friend and foe in the thick jungle canopy. The Americans suffered a staggering 70 percent casualty rate. A beaming Kinnard reported later that the "battalion had taken everything the enemy could throw at it, and had turned on him and had smashed and defeated him." The 3rd Brigade handed off operations to the 2nd Brigade, and the campaign continued until November 26, when authorities denied Kinnard's plea to pursue PAVN forces into Cambodia—the president had no intention of widening the war. Kinnard viewed the fight as a traditional cavalry pursuit situation. "Not to follow them into Cambodia," he said in 1990, "violated every principle of warfare."[54]

To the Americans, the battle showed airmobility as a viable, if tenuous, concept. After combat in the Ia Drang, Kinnard declared that the infantryman was finally "freed from the tyranny of terrain" thanks to the helicopter. Helicopters had helped Moore's battalion achieve surprise. They helped sustain the fight by providing logistical support and medical evacuation, albeit only by the heroic efforts of the pilots involved in these missions—one of whom earned a Medal of Honor for his actions at LZ X-Ray. Helicopters also helped sustain American firepower—critical to both fights, especially McDade's battalion at Albany—by delivering much-needed ammunition. Further, helicopter gunships and aerial rocket artillery were influential during the battle. A Central Intelligence Agency report called it the "greatest success of the war" to date. Much of the chatter after the fight centered on how it proved the validity of airmobility and that it demonstrated that main force North Vietnamese units would remain ineffective against American forces in a conventional battle.[55]

Nothing could be further from the truth, as the debacle at LZ Albany illustrates. Moreover, success at LZ X-Ray came despite Moore's battalion's inability to maneuver as airmobile doctrine called for. The Vietnamese initiated contact and withdrew when they wanted. They learned to counter helicopter-mounted forces by avoiding large units, setting ambushes, and fighting so close to American units that the American air and firepower advantage could not help. Helicopters were loud and flew relatively low and slow, thus providing an opportunity for Vietnamese forces to ambush Americans after learning from the Pleiku campaign. Airmobile operations were also dependent on the presence of suitable landing zones. Any obvious helicopter landing spot would present equally obvious ambush locations in areas with limited LZ space. Moreover, despite what Westmoreland called an "unprecedented victory," North Vietnamese infiltration rates did not decrease, Hanoi seemed unfazed, and North Vietnamese forces demonstrated that they would be the ones to choose the time and place of battle. One adviser noted the irony of seeking success in helicopter warfare, stating that "when you come to think of it, the use of helicopters is a tacit admission that we don't control the ground. And in the long run, it's control of the ground that wins or loses wars."[56]

The helicopter provided mixed results. Even the army chief of staff had doubts about the cavalry's performance in Ia Drang. The battle set the standard for measuring effectiveness based on body count; it established "a hierarchy among metrics most important to the chain of command." Many units, including the 1st Cavalry Division, adapted their tactics afterward, focusing on small-unit movements and long-range foot patrols, abandoning the quick airmobile dispersal and massing concepts to instead use helicopters as mere transportation for moving around the mountainous and jungle terrain in South Vietnam. These

FIGURE 6.   The 4th Battalion, 503rd Infantry, 173rd Airborne Brigade prepares to be heli-lifted by UH-1D helicopters to the brigade's forward base camp in Xuan Loc Province after completing a search-and-destroy mission. It was the last flight of UH-1D helicopters to land and offload the remainder of the unit at the brigade's forward base camp, on September 1, 1966. US Army photo, RG 111, NARA.

tactics buttressed increased emphasis on civic action and rural construction programs throughout the 1st Cavalry Division's time in the Central Highlands. Nevertheless, helicopters were integral to how the army prosecuted the war in Vietnam. Continued innovation and adaptation helped carve out an essential role for heliborne infantry formations to maneuver above the unforgiving jungle terrain of Vietnam while providing an essential vehicle for resupply and casualty evacuation. Though imperfect, the 1st Cavalry Division's actions in the Pleiku campaign put Gavin's concepts into action and inaugurated the "helicopter war."[57]

The UH-1 Huey became one of the preeminent symbols of the American war in Vietnam. And rightfully so: in July 1969, the US Army had approximately thirty-five hundred helicopters "in country," with about 72 percent operational on a given day. Gen. Westmoreland called air-mobile warfare "the most innovative tactical development to emerge from the Vietnam War." This was despite the enemy's ability to learn how to avoid massive air assaults and thus avoid fighting large units, which forced American units to change their tactics and adopt long-range patrolling techniques to find, fix, and destroy PAVN and the People's Liberation Armed Forces (Vietcong). It is also important to remember that airmobility was not developed for the Vietnam War but rather for the requirements of a perceived nuclear conflict with the Soviet Union. Nevertheless, airmobility was not a technological stopgap that helped the army avoid counterinsurgency, as Andrew Krepinevich argues, but rather a tool born of an ingrained cultural predilection toward air-mindedness in a subset of the army—the airborne mafia. Helicopters—whether designed for assault, resupply, reconnaissance, or other missions—defined how the US Army fought in Southeast Asia.[58]

For three years in Vietnam, the army fielded two airmobile divisions, as the 101st Airborne converted to an air assault division in 1968 after army leaders realized the futility of maintaining qualified paratroopers in a combat zone. A minor uproar ensued about what to call the division; the Pentagon ordered it called the 101st Air Cavalry Division, then the 101st Infantry Division (Airmobile) before former 101st commander William Westmoreland took over as chief of staff and directed that the two airmobile divisions retain their historic designations as "airborne" and "cavalry." With no opportunity to maintain proficiency in parachute training and an influx of draftees without airborne training, army leaders decided to staff nominally "airborne" units with non-airborne personnel to keep replacement rates sustainable. The 1st Cavalry Division was also caught in this minor fiasco. Nevertheless, the reorganization took a year as the Screaming Eagles continued combat operations and had to procure the required helicopters. Of course, the commander to lead this transition was another World War II–era paratrooper, Maj. Gen. Melvin Zais. Effective April 1, 1974, the 101st's final airborne brigade was deactivated, and the division was entirely airmobile.[59]

As the American experience in Vietnam gave way to late Cold War restructuring, the 1st Cavalry Division was redeployed to Fort Hood and reorganized as a "triple capability" (TRICAP) division. The new formation combined mechanized and armored forces with an air cavalry combat brigade consisting of airmobile infantry that resembled the Howze Board's air cavalry brigade concept yet incorporated lessons learned in Vietnam. This design was short-lived, however, as the 1973 Yom Kippur War galvanized the army into adding more heavy tank units to its force structure. In 1975, the army separated most aviation assets from the 1st Cavalry Division to create the 6th Air Cavalry Brigade, rendering the 1st Cavalry Division an exclusively armored unit. This returned the army to a one-air-assault and one-airborne division force that was rapidly mechanizing to fight across the plains of Europe, relegating to the back burner the light, highly mobile forces poised to respond to brushfire-type wars. For example, the 1978 twenty-four division, twenty-four brigade total force consisted of just 3 light infantry, 9 air assault, 10 airborne, and 88 standard infantry battalions, yet boasted 108 mechanized and 129 armored battalions.[60]

Regardless, developing airmobility paid dividends in restoring an airmindedness to the army not seen since before the air force achieved independence in 1947. Furthermore, the organic nature of the airmobile division's helicopter assets allowed a new culture to develop, because any combination of organizations working together over a period will develop shared values, beliefs, and norms about the way they interpret doctrine and therefore how they operate through tactics, techniques, and procedures. This does not happen with ad hoc organizations that rely on doctrine without familiarity. Despite its rapid mechanization, the US Army maintained a strong airmobile capability—each division now has its own aviation brigade—that paid dividends in Grenada, Panama, the Persian Gulf, and the twenty-first-century invasions of Afghanistan and Iraq. The ability to fly over the environment, seize key terrain, screen, provide flank protection, and delay large enemy formations while delivering infantrymen about the battlefield proved instrumental in the army's post–Cold War operations.

CHAPTER 6

The Strategic Army Corps and the Emergence of Strike Command

> The nation or group of nations that control the air will control the peace.
>
> —James M. Gavin, *Airborne Warfare*, 1947

The mushroom clouds that rose over Hiroshima and Nagasaki in August 1945 ushered in a new era with an emphasis on atomic weapons and aerial warfare. The ability to deliver a devastating payload from the sky shifted the brand-new United States Air Force to the forefront of the nation's military strategy. The primacy of the air force demonstrated the country's continued infatuation with the promise of technology and industry to win wars without bloody ground warfare. The future was in the air, and Gavin and many other army officers understood this well—a second product of their relative air-mindedness from past experiences. "This means," Gavin wrote, "being able to transport airborne troops to any spot on the globe. It means being able to deliver those troops, trained and equipped, and capable of imposing their will on any potential or actual belligerent." The airborne mafia, leaning on World War II experiences and a predilection for air-mindedness, provided the necessary leadership and background to develop and maintain rapid-response forces imbued with an expeditionary mindset.[1]

The quest for an entirely air-transportable army ensued. In 1947, Gen. Jacob Devers wrote that when the United States could develop "an Army which can fly to fight and an Air Force that can fly it, there is every prospect that we can avoid war and make peace permanent. The guarantee of sure, swift retribution will certainly deter any would-be aggressor." Gavin wrote that future war would place a "predominant emphasis on airpower," which would give airborne forces—those with a natural predisposition to air movement—a crucial position in any future military. "The troop carrier pilots [and] the airborne troopers, will play the leading role in future aerial combat," he wrote. "Together they provide the means of delivering a decisive blow anywhere within the capabilities of the aircraft they employ." Mobility and rapid force projection were the keys to future warfare, and air-minded, airborne forces were perfect for executing this type of warfare. "Mobility" is an ambiguous term but comes in two forms—strategic and tactical. Strategic mobility is the ability to move forces *to* the battlefield, while tactical mobility refers to moving forces *on* the battlefield. Because of their air-minded expeditionary mentality, the airborne mafia was well suited to precipitate the change required to keep the army relevant during this era.[2]

Airborne units, therefore, provided the army with a strategic asset. They were readily adaptable to respond to multiple contingencies. Contingency responses are inherently expeditionary, joint, and ideally short in duration to preserve the contingency force for future operations. Contingency forces like those the airborne pioneered in the postwar period offered American strategic planners a fire brigade to respond to varied crises around the world. This capability, as air force transport capacity increased, allowed the US armed forces to project ground combat power from the United States. As massive retaliation relegated ground forces to second-tier status, honing a contingency way of war predicated on rapid response helped provide the army with increased legitimacy.[3]

Thanks to the efforts of the airborne mafia, an emphasis on air-mindedness permeated the army by the late 1950s. The pentomic division was intended to be mostly air transportable, and helicopters were in development. Increased strategic mobility capabilities gave innovative airborne thinkers the chance to experiment and grow. The army's airborne community kept it relevant through its foresight and reinvention as a strategic response force. The army developed reaction forces such as the 1953 designation of the 82nd Airborne Division as the Western Hemisphere Reaction Force, the designation of the XVIII Airborne Corps as the Strategic Army Corps (STRAC) in 1958, and a new joint headquarters, US Strike Command (STRICOM) in 1962.[4] These ideas stem from historical experiences reinforcing Allied forces in combat at Salerno and Bastogne during World War II. The expeditionary mindset that developed in World War II helped the airborne mafia redefine the army's role by emphasizing strategic mobility and faith in delivering land combat power by air to the battlefield. Rapid-response capabilities became a critical component of postwar planning, and amid a national focus on airpower, strategic response forces helped bridge the gap between the army's personnel reduction and its military mission of conducting land warfare against enemy armies. The experience of World War II led army leaders to develop capabilities for rapid aerial responses to a wide range of contingencies. This helped the army maintain its relevance in an era of nuclear-based air force dominance by providing policymakers with alternatives to nuclear warfare in crises that may not have required such a drastic response.

Toward an Airborne Army

Following World War II, many influential officers envisioned what they referred to as an entirely airborne army. The belief was that airborne operations would play a significant role in the future and that the United States needed the capability to deploy complete major combat units by air. A study recommended that every piece of equipment in the army's inventory be lightweight and air transportable—not necessarily for the entire army at once, as the service would still move most of its forces by sea, as it does today; but every item of equipment would at least meet stringent air force requirements for aerial transport. The wartime commander of the First Allied Airborne Army, Lt. Gen. Lewis H. Brereton (a rated pilot) advocated for a large peacetime self-sustaining and self-supporting airborne organization operating directly under the Department of Defense. Brereton's Airborne Army was created before Operation Market Garden to provide command and control for all Allied airborne units for the remainder of the war. It was Ridgway's higher headquarters and also controlled the troop carrier units. Gen. Omar Bradley noted the importance of lightweight, air-transportable equipment when he addressed the 1949 graduates of the Command and General Staff College. Yet by the end of the decade, a typical infantry division of seventeen thousand men would require 176 C-82 Packet and 88 C-54 Skymaster transport aircraft, operating continuously, to move the entire unit in twenty-nine days. Creating an air-transportable force that would give the entire service an expeditionary capability and match the era's infatuation with air power was a pipe dream.[5]

The Cold War and the positioning of the Red Army near US allies in Europe meant the United States would no longer have months or years to build military forces. The army needed combat units to reach the battlefield within hours instead of weeks. Permanent readiness and the ability to respond to multiple crises led to the creation of a ready reserve in the United States. The army designated its lone post–World War II airborne division as a strategic striking force. This idea was less about parachuting into combat but rather about rapid air transport. According to Taylor, "the airborne concept to me is the capability of rapid movement by air of military units, both for tactical and strategic purposes." At the outbreak of hostilities in Korea in June 1950, the army considered the 82nd Airborne Division its only combat division ready to fight—the rest of the army was either on occupation duty or missing much of its allotted strength. The division was alerted for potential employment in Korea, though never sent. There was still insufficient transport aircraft to lift even half of an airborne division in 1950. This situation did not improve until the C-130 was developed in the late 1950s.[6]

Strategic mobility intrigued leaders in the postwar army as they reevaluated the requirements for the size of the airborne division. As constructed in World War II, the 82nd Airborne Division had little firepower compared to the airborne division of 1950 that included a tank company—identical to the standard infantry division. The mission of the parachute infantry remained essentially the same since the 1943 issuance of Training Circular Number 113—airborne troops were to be committed by air transport only on missions that other forces could not more expeditiously perform. As airborne forces were not to be employed unless other forces could support them within three days, the general assumption was that they would be used as part of a larger operation. In the future, however, airborne forces needed more firepower and air-transportable equipment—an almost impossible combination with existing air transport capabilities—to increase their effectiveness once on the ground. Increased effectiveness in ground combat would earn aerial-delivered forces greater autonomy.[7]

Bradley envisioned the next war in three stages. First, the United States would use strategic nuclear weapons. Next, the United States would seize bases either near or within the enemy's homeland. Finally, a large-scale ground assault would defeat the enemy. Airborne forces would be a key component in phases two and three. Six months later, Gen. Ridgway announced the army's plans "to place increasing emphasis upon airborne, air-transportability and air-ground support techniques." In a February 1950 memorandum to Maj. Gen. William Miley, Lt. Col. Melvin Zais wrote that the future concept of airborne operations must include provision for airfield-seizure-type missions. Zais, an airborne battalion commander in World War II, echoed the postwar General Board for United States Forces in the European Theater, which recommended maintaining airborne divisions for forced-entry situations while emphasizing that the air transport of an infantry division would become a vital mission in future warfare. The board found that there was no acceptable alternative to the airborne division.[8]

Landing troops in slow-moving transport aircraft required a secure airfield. Seizing an airfield in enemy-held territory required trained parachute forces to accomplish the feat without landing airplanes. However, seizing, reinforcing, and expanding the perimeter of an airfield required massive amounts of airlifted supplies, something the fledgling US Air Force was not yet ready to provide. Plans in 1952 called for the 82nd Airborne Division and one marine division to be prepared to fight in Europe thirty days after mobilization. Should the Soviet Union invade Western Europe, airborne forces were needed to regain a foothold on the Continent. The use of parachute forces combined with atomic weapons featured prominently in Ridgway's plans for reinforcing Europe. When discussing exercise priorities with his successor as commander of Supreme Headquarters Allied Powers Europe, Gen. Alfred Gruenther, he stressed that "of far greater importance to us, in my opinion, will be a capability to move and strike rapidly by air, in conjunction with the use of special [atomic] weapons." He urged Gruenther to stress joint army and air force training because he believed that "the ability to move across large bodies of water and reestablish a lodgment in Europe through the combined use of special weapons and airborne assault forces may well be decisive."[9]

Nevertheless, the army and its mission of reinforcing Western Europe in the event of a Soviet invasion suffered because of the air force's steadfast resistance to developing more transport aircraft. This resistance was based on the institutional preference for more and better bomber and fighter aircraft to fulfill the nuclear delivery and air superiority missions. This resistance continued even though the Berlin airlift demonstrated to military leaders the value of air-landed resupply during contingency operations when no other recourse was possible. In a defining moment for the new US Air Force, the operation's zenith had an aircraft landing in Berlin every sixty-two seconds. By delivering eight thousand tons or more of cargo daily, the operation proved that the air force could deliver large quantities of supplies with its current transport fleet. It was, therefore, easy to draw the connection between the ability to feed 2.5 million people and the ability to deploy and support a ground combat force by air alone. Yet despite the herculean effort of the fliers, they did not provide sufficient calories, and Berliners often turned to the black market to get what they needed. Regardless, Lt. Col. William Kuhn—an airborne battalion commander wounded in Normandy—used the Berlin airlift as an example in his article on airborne armies. He noted that planners in 1950 were no longer thinking of only World War II–style airborne units but entire air-transported armies. "The planners of today are thinking in terms of airheads established by airborne and air-transported corps equipped with air-transportable howitzers, tanks, bulldozers, and other essential weapons and vehicles," he wrote. Gen. Devers and other army leaders considered the Berlin airlift a critical turning point toward creating fully air-transportable army divisions. It was the proof the army needed of the air force's ability to insert land combat forces into airfields behind enemy lines and keep them supplied.[10]

By 1950, the air force could still not move a two-division corps to Europe in fewer than ninety days. The airlift capability to do so did not exist. In May 1950, Ridgway reported to the army vice chief of staff that the army could not meet its mission of reinforcing Europe with two divisions in fifteen days if a crisis arose. At best, it would have taken seventy days. Brig. Gen. Lemuel Mathewson, commander of the 11th Airborne Division at the outbreak of the Korean War, lamented the lack of aircraft and complained to Ridgway that it was time for the air force not only to rebuild its troop carrier fleet but also to give the transport pilots more prestige. Ridgway was an ardent supporter of airpower but saw an overemphasis on one form—the long-range bomber—at the expense of others. To Ridgway, the army of the future would be dependent on aircraft, and the air force needed the capability "to lift whole armies, armed with nuclear weapons, and put them down upon any spot on the earth's surface where their tremendous, and selective, firepower will be needed." The financial cost to maintain both a large Strategic Air Command and the troop carriers required for airborne and air-transportable troops was not feasible within the budget constraints before the summer of 1950.[11]

In April and May 1950, a multidivision airdrop training exercise at Fort Bragg, North Carolina, known as Exercise Swarmer, confirmed the problem of aircraft availability. The field maneuver was the first attempt to insert an airborne force into an airfield and sustain it until follow-on forces arrived. In this exercise, the entire air force troop carrier fleet could drop only one regiment at a time, and the drops were scattered over multiple days. Yet Swarmer was not without its successes. It was the first exercise to feature large-scale heavy drop operations. The relatively new C-119 "Flying Boxcar" transport aircraft represented a significant improvement over previous transport aircraft. Before its development, the heaviest droppable item was a 75mm howitzer, which could only be dropped in multiple pieces—making putting the gun into action on the drop zone an adventure. The C-119 could transport forty fully equipped paratroopers or sixty air-landed troops. Its increased payload allowed it to deliver items as heavy as a medium-size bulldozer through its rear clamshell-style doors. The critical component of the C-119 was its floor-mounted monorail system, running through the center of the aircraft. This system allowed for the ejection of bundles from a forward hatch before paratroopers exited the sides. More importantly, this monorail allowed large items to be dropped from the rear in a controlled fashion. During Exercise Swarmer, approximately 85 percent of all airdropped supplies were recovered with no damage, and thirty of thirty one-quarter-ton trucks were operational after being dropped—a more than twofold increase over World War II experience. This exercise also included four 105mm howitzers dropped intact, thanks to the monorail system. Despite the lack of aircraft, Swarmer served as a preview of future operations in Korea and led to the creation of a joint board to solve problems discovered during the exercise.[12]

Putting the lessons of Swarmer into practice, 1,470 paratroopers of the 1st and 3rd Battalions of the 187th Airborne Regimental Combat Team (RCT) exited their C-119 and C-47 aircraft over Sukch'on, Korea, on October 20, 1950. Simultaneously, the 2nd Battalion jumped fifteen miles east at Sunch'on. Minutes before the personnel drop, for the first time in combat, jeeps, antitank guns, howitzers, and the requisite ammunition fell from the sky under nylon parachutes. Seventy-one C-119s and forty C-47s delivered 2,860 paratroopers and over three hundred tons of supplies in a few hours. On the first day, twelve howitzers, four antitank guns, their requisite vehicles, ammunition, and a host of other supplies found their way to the drop zone ahead of Eighth Army forces near Pyongyang. Between October 21 and 23, forty C-119s delivered more personnel, howitzers, and equipment to the troopers on the drop zones. The 187th RCT performed well, capturing 3,443 and killing more than

FIGURE 7.   US Air Force Fairchild C-119 Flying Boxcar transport planes of the 403rd Troop Carrier Wing, 315th Air Division, spill out their load of heavy equipment for the men of the US Army's 187th Regimental Combat Team during a maneuver in Korea on October 1, 1952. US Air Force photo.

450 of the more than 5,000 North Korean soldiers it encountered. The operations in Korea involved the heaviest equipment dropped in combat to that point. Because of the difficult terrain, heavy supply drops became standard for all units during the Korean War, made possible by the C-119's ability to transport much heavier equipment.[13]

A Joint Airborne Troop Board, established on April 26, 1951, met multiple times over the next six years and held three major conferences in 1951, 1953, and 1957. The board was charged with studying new concepts for the application of airborne units in future warfare while laboring to improve interoperability. Its initial goals were to develop a suitable concept for operations, estimate the army's requirements for airborne operations in the next war, and recommend programs for training in peacetime to maintain readiness. An additional task was to assess the effectiveness of equipment and organizations. The initial iteration of the board declared the venerable towed glider obsolete. Instead, the board called for helicopters, small assault transport planes, and "convertiplanes" to replace its payload delivery capabilities. A convertiplane is an aircraft that uses rotor power for vertical takeoff and landing and converts to fixed-wing lift for normal flight, as exemplified by the Boeing V-22 Osprey and the new Bell V-280. Two ideas were developed, the Bell XV-3 and the McDonnell XV-1, but neither entered production.[14]

The Joint Airborne Troop Board's findings and recommendations laid the groundwork for developing air mobility throughout the decade. One of the panel's main conclusions was the need to procure improved troop carrier airplanes in larger numbers, something air force bomber-oriented generals resisted. More importantly, the board established joint doctrine for airborne operations, noting that "there must be a unified airborne, striking force in being with the necessary airlift earmarked and capable of immediate mobile deployment," a crucial step in the development of joint air mobility. The result was the publication of the Joint Action Armed Forces manual, published in the Army as Field Manual 110–5, Navy JAAF, and Air Force Manual 1–1.[15]

In a series of 1953 speeches, the under secretary of the army, Earl Johnson, stressed the need for greater mobility to offset Soviet numerical superiority and to move land forces quickly across oceans to meet threats. His August speech to the Air Force Association's national convention was sent directly to the Joint Airborne Troop Board, which was then studying mobility concepts at Fort Bragg, and reprinted in the December 1953 issue of *Army Information Digest*. To the army, achieving full air transportability was paramount for preparing for the next war. "This problem of becoming airborne and mobile is not just an exercise," Johnson said. "It is a considered goal which we must not fail to attain. If war should come, an airlift capacity must exist which is, on the one hand, transoceanic so that we can deploy rapidly, and which, on the other hand, is capable of lightning quick assault." Johnson acknowledged the expensive nature of airlift capabilities but reminded the air force of its role and the army's efforts to streamline equipment for air transport. He closed his speech by calling on the air force to serve as the army's all-too-important partner in achieving his service's goals. "Airpower properly fitted to Army needs," said Johnson, "greatly enhances the Army's mobility." Later in November, Johnson alluded to the future of air mobility when he remarked that "fortunately for the Army and the country, the new Secretary of the Army, Robert T. Stevens, is fully air-mobile-minded." The aerial movement of combat troops was poised to become the primary means of moving combat forces worldwide.[16]

Nearly every exercise throughout the 1950s featured airborne operations, which represented the level of institutionalization that airborne units had reached. This was despite the realization that the parachute was a "very inefficient means of transport." The need to develop more means of inserting forces was necessary. In October 1955, Taylor stressed that "the Army today is bent on reaching a condition of airborne effectiveness which we are convinced is necessary for the successful conduct of modern warfare. We want an Army with many completely air-transportable combat units, complete with sufficient weapons, vehicles, and supplies to sustain themselves in ground operations." The goal was not just paratroopers but fully air-landed units capable of fighting upon arrival.[17]

Nevertheless, the army relied on airborne units because of those units' familiarity with aerial transport. In 1956, the army transported an entire regiment of paratroopers from Fort Campbell in Kentucky to Japan and then back to Fort Bragg, North Carolina, in ten days to showcase its strategic air mobility. Airborne units had gained such prestige that regular infantry commanders called for a curtailment of parachute training as a reenlistment incentive since it diverted much of the army's manpower to these units. By Ridgway's retirement in 1955, the army had a "mobile ready force" under the guise of a strategic reserve in published policy. Still, it was inadequate and was competing with such budget black holes as nuclear weapons and continental defense requirements for funding. Nevertheless, in his retirement letter to the secretary of the army, Ridgway called for funding a joint mobile force of such versatility that it could apply to almost any scenario, sowing the seeds for what became STRAC.[18]

In 1957, the assistant deputy chief of staff for military operations, Maj. Gen. Earle Wheeler, noted that the current long-range airlift option, the C-124 Globemaster II, made transporting one airborne division in three hundred sorties possible. However, this was without logistical support—only the division's combat echelons could deploy with this number of sorties. Wheeler also saw that the army's ability to win and deter general or local wars depended on strategic mobility. Heavy drop and the ability to land on unimproved runways allowed more combat power to reach the objective area in less time. The double-decker configuration of the C-124 was imperfect for parachute operations but allowed for robust cargo capabilities. The replacement of the C-119 with the new C-130A Hercules in 1956 doubled lift capacity, as it could carry twice as much tonnage or personnel. When the upgraded C-130B entered operational service in 1958, the army's strategic reach doubled thanks to the aircraft's increased range.[19]

According to Maxwell Taylor, strategic mobility was the key to developing "a powerful Army capable of coping promptly with military situations wherever they may occur. . . . It is the combination of firepower and mobility that wins war." Army efforts in touting its airborne capabilities were not wasted, and by June 1957 even Secretary of Defense Wilson included paratroopers alongside such strategic assets as bombers and missiles in a speech, a sign of the airborne's importance as a national security option and credible deterrent. While the service had become increasingly expeditionary since 1898, the airborne mafia successfully imparted their expeditionary vision to the rest of the service and ensured the entire Department of Defense heard their pleas.[20]

The Strategic Army Corps

Airborne forces were rapidly becoming the army's primary component for deterrence in the Eisenhower era. They featured in every major exercise, and with Taylor and Gavin directing the service's priorities, they became the veritable military poster children of the age. The next logical step was an official organizational structure for a large rapid-response force. This would soon evolve into the Strategic Army Corps. The concept was initiated in 1957 but not formally announced until 1958. STRAC received "maximum emphasis to prime its units for immediate movement into combat anywhere in the world." The idea was to create a force capable of responding to any global crisis, from general war to irregular conflict. Gavin wrote in 1955 of the need for a dual-capable force that could "mobilize for large-scale war" while having "sizable forces in being ready to move rapidly and put out 'brush fires' before they get out of control." This type of force, he believed, would provide a credible deterrent to adversaries. STRAC, then, became a four-division combat-ready force that was "ready to move at a moment's notice," as one article described its posture. STRAC was composed of the 82nd and the 101st Airborne Divisions, as well as the 1st and 4th Infantry Divisions. At least one of the airborne divisions was on alert to fly into a hot spot on short notice. STRAC was primed to demonstrate the efficacy of strategic mobility in action.[21]

Needing a mobile corps-level command element, the army reactivated the XVIII Airborne Corps to provide necessary command and control. As the core of the broader nine-division Strategic Army Force, STRAC could give American leadership more than 125,000 troops ready to respond to global crises. The corps fell under the US Continental Army Command for training but would be turned over to the established command structure in the theater of operations if activated. Its primary mission was to reinforce American or Allied forces in Europe or the Far East, and at least one of its airborne divisions had previously, since 1953, been prepared to move on a moment's notice as a Western Hemisphere response force. Readiness became STRAC's watchword as the corps resembled the US Air Force's Strategic Air Command, at least on paper. In fact, Taylor admired SAC and hoped to replicate in the army its dedication to readiness. The STRAC motto—"Skilled, Tough, Ready Around the Clock"—took on a life of its own throughout the army, as a "STRAC trooper" referred to any squared-away soldier like those on alert. Lt. John R. Galvin capitalized on this acronym to galvanize morale in his beleaguered administrative section. Also, while airborne divisions comprised parachute-qualified volunteers, they additionally enjoyed a unique ability to transfer marginal, low-performing soldiers out of their ranks, thus reinforcing their brand of discipline and readiness.[22]

In practice, however, the concept was imperfect. STRAC's state of readiness was often in doubt. Mass parachute drops accompanied nearly every major exercise, but these often served as showpieces—better suited to public relations opportunities than combat. Many of the paratroopers, it seemed to Maj. Gen. Hamilton Howze, the 82nd Airborne Division commander in 1958, were interested only in the jump and disdained anything that came after. He was not the only airborne officer to comment on training deficiencies affecting morale during the decade; Robert Haldane, who commanded a battalion in the 82nd in 1959, rarely had more than eighty paratroopers in each company and observed that the 82nd was "never really trained as well as they thought they were. There were too many distractions." In addition to personnel and equipment shortages, commands had poorly developed standing operating procedures governing movement, and procedures that did exist were often not synchronized across the rest of the corps.[23]

Herbert Norman Schwarzkopf, the future commander of coalition forces in the 1991 Gulf War, served as a young platoon leader in the 101st Airborne Division at Fort Campbell from 1957 to 1959. As a brand-new lieutenant, he explicitly requested assignment to the 101st "because it was part of the vanguard, the Strategic Army Corps, and had lately attracted publicity as America's first 'pentomic' division, specially tailored to fight on the atomic battlefield. It had a magnificent tradition as well." Schwarzkopf loved the glamour and mystique of the 101st but quickly grew disillusioned at the number of untalented leaders who rested on the laurels of the unit's history. "The Strategic Army Corps trumpeted itself as a great fighting force," he wrote, "but we knew we really weren't that good. We could see it in our officers . . . and the quality of our equipment." Distractions like demonstration jumps and a shortage of air force aircraft exacerbated problems, limiting the STRAC divisions to only a few significant annual exercises. While Schwarzkopf's experience highlights some of the issues affecting the pentomic division, most airborne units assigned to STRAC maintained a better than 90 percent readiness rating. In comparison, the infantry divisions hovered around 70 percent.[24]

Rather than moving to deter a Soviet onslaught, STRAC's first major employment came within the United States. In September 1957, at President Eisenhower's request, a battle group of the 101st Airborne Division was mobilized in response to rioting in Little Rock, Arkansas, over the admission of African American students to a high school in the city. The paratroopers were selected because they were already on alert and ready to move, and Taylor wanted to demonstrate the army's quick-response capability. The 1st Airborne Battle Group, 327th Infantry, was deployed in four trips between their home base at Fort Campbell, and Little Rock. The first aircraft landed at Little Rock Air Force Base a mere four hours after the battle group received its initial alert, while the entire battle group was on the ground at Little Rock within nine hours on September 24. The officer placed in charge of the Arkansas Military District, Maj. Gen. Edwin Walker, on his way back from a Pentagon briefing, stopped at Fort Campbell and briefed the 101st Airborne commander, Maj. Gen. Thomas L. Sherburne Jr., on the situation in Little Rock. Walker, a staunch white supremacist, advised Sherburne to "reduce, very discreetly, the 'colored strength' of his task force; and to ensure that black infantrymen who went to Little Rock were out of direct contact with the public." The paratroopers personally escorted Black students into Central High School and were a welcome change, as the agitators in the situation did not respect the authority of the National Guardsmen, who had helped prevent desegregation until federalized by Eisenhower. The paratroopers dispersed the mob and returned to Fort Campbell by November 27 after the situation stabilized.[25]

The following year, on May 13, 1958, when Vice President Richard Nixon's motorcade came under attack from rioters in Venezuela, the president and Joint Chiefs positioned military units to respond quickly if needed in the aftermath. This included STRAC personnel from Fort Campbell, US Marines at Guantanamo Bay, Cuba, and the aircraft carrier USS *Tarawa*. The attack represented a culmination of widespread opposition to Nixon's eight-nation tour of Latin America. After stops in Uruguay, Argentina, Paraguay, and Bolivia, the Nixon entourage experienced an inconsequential rock-throwing crowd in Peru. Tensions seemed to cool as Nixon visited with leaders in Ecuador and Columbia, but in Venezuela, the vice president's group faced increasing tensions, culminating in multiple blockades by several hundred protesters, who shattered the windshield of Nixon's vehicle. The motorcade made it through the riot, assisted by a group of reporters, but this proved to be one of the scariest moments of Nixon's life. News of the event reached the United States quickly, and STRAC had its first overseas mission.[26]

Just before 4 p.m., the telephone rang at the 101st Airborne Division commander's office at Fort Campbell. In little less than three hours, two rifle companies of paratroopers—about four hundred men—were on their way to Puerto Rico on twenty-two C-130s. In less than five and a half hours, they were seventeen hundred miles from home, staged at Ramey Air Force Base, in Puerto Rico, awaiting further instruction. The task force, commanded by Col. Robert C. Works, was from the 1st Airborne Battle Group, 506th Infantry, and had enough provisions for five days of fighting. Not expecting a parachute drop, the men were nonetheless prepared to land and fight if needed. The 101st Airborne Division's alert and movement, according to then-STRAC and XVIII Airborne Corps commander Lt. Gen. Robert Sink, who had commanded the 506th Parachute Infantry Regiment in World War II, "demonstrated the ability to move into action rapidly and decisively to put out a brush fire anywhere before it can develop into general war." The rioting relented; US troops did not land on Venezuelan soil, and the vice president made it home unscathed. The 101st troopers returned to Fort Campbell on May 15.[27]

The episode in Venezuela demonstrated the potential of STRAC. Following the incident, President Eisenhower suggested carrying out exercises and airlift operations outside the continental United States so that when the US did respond to contingencies, the troop movement would not cause alarm. While intended to reinforce Allied forces in Europe in the event of a Soviet invasion, the STRAC-ready brigade concept quickly proved ideal for responding to global crises, putting out "brushfires" before they turned into something larger. By the middle of the decade, however, officers realized that using quick-strike, lightly armed airborne forces against mechanized armed forces of an industrialized nation, and in the era of surface-to-air missiles, was akin to suicide. Nevertheless, the value of these forces in advancing US foreign policy in developing nations while protecting American interests and citizens was on full display in the late 1950s. Moreover, having a force "ready around the clock" provided the United States with credible combat power prepared to project anywhere in the world and deter would-be challengers.[28]

After the Venezuelan incident, the chairman of the Joint Chiefs of Staff called for a reappraisal of emergency airlift plans. In the ensuing report, the Department of Defense boasted two dedicated airborne quick-response capabilities. The Tactical Air Command (TAC) and Continental Army Command (CONARC) maintained a task force of two 82nd Airborne Division battle groups and fifty-two C-130 transports ready to move within twenty-four hours of notification. A second plan under development relied on the Military Air Transport Service to move two battle groups of the 101st Airborne Division to a secure airfield in Europe or the Middle East. That plan required eight days and 361 trips to move five thousand personnel and 4,500 tons of supplies. Expansion to include the rest of the division would require eighteen days and nine hundred further sorties to move the rest of the 101st's 15,000 tons of personnel and supplies.[29]

Airborne response forces faced another test in the summer of 1958 when Lebanese president Camille Chamoun requested assistance from Egyptian leader Gamal Abdel Nasser in deterring Communist subversion. Ostensibly acting to maintain American credibility, Eisenhower reluctantly sent American troops to the region. The initial plan called for the 101st Airborne and 4th Infantry Division to fly directly into

Lebanon as part of STRAC in a move known as Operation Swaggerstick. A War College study determined that deploying the entire STRAC to Lebanon would require more than three-quarters of the US Air Force's available airlift assets. Rather than deploy STRAC, the Joint Chiefs elected to send a US Marine Corps task force and a reinforced airborne battle group from Germany so as not to commit the entirety of air force lift assets at once. Initial plans called for the airborne battle group to secure a lodgment to allow British follow-on forces to fly in and assist, if met with a hostile environment. The Joint Chiefs clamored for a parachute drop throughout the planning phase, even clearing the necessary airspace. Meanwhile, the airborne battle group from Augsburg, Germany, flew to Adana, Turkey, before landing in Beirut. This force consisted of the recently reorganized 24th Airborne Brigade, a semiofficial designation for the two airborne battle groups of the 24th Infantry Division that had assumed the airborne mission in Europe after the army deactivated the 11th Airborne Division. American forces under the command of Maj. Gen. Paul D. Adams secured Beirut's airport, seaport, and the road approaches leading into the city. Throughout the crisis, the 101st Airborne Division remained on alert, ready to send its initial elements within six hours of notification.[30]

Operation Blue Bat, the name of the Lebanese operation, validated the concept of airborne elements as a quick-reaction force by, in this instance, landing the paratroopers in 110 C-130s and C-124s. However, the operation was not without its problems, and it further illustrated the challenges the air force faced in providing enough airlift to support the army's plans for strategic mobility. The airborne troopers in Lebanon exhibited high morale and impressed the task force's commander despite their tenuous mission as essentially a show of force, and to support President Chamoun amid civil strife. The army claimed it needed the air force to maintain forty-five troop carrier wings of between thirty-six or more aircraft, or enough to carry the assault forces from three airborne divisions simultaneously—a number the air force was not remotely interested in procuring. Despite STRAC's role as an easily deployable force for emergencies, aircraft limitations prevented it from truly living up to its potential. Without a declared national emergency, procuring the required lift assets to support a full-scale troop deployment was impossible. Blue Bat also demonstrated that the air force had been neglecting its airlift capacity and reinforced the notion that conventional ground forces remained important in the atomic era. Ultimately, the Lebanon experience demonstrated that the entire

Department of Defense needed to be better prepared for limited overseas operations and interventions.[31]

By 1959, STRAC had lost one infantry division to the mission of initial recruit training, making it a three-division corps. Gen. Lyman Lemnitzer, who succeeded Taylor as army chief of staff, emphasized at the Sixth Annual Conference of Civilian Aides to the Secretary of the Army that with two-thirds of its divisions being airborne, STRAC was capable of employment in areas that lacked "ports or airfields or even beaches—necessary for other types of forces to be landed." The airborne force "could parachute into a zone where combat was actually taking place, and seize airfields, beaches, or ports to defeat the enemy or permit the landing of other forces." Later that day, Lt. Gen. Sink, the STRAC commander, outlined the specific capabilities of his corps. In addition to the 82nd, 101st, and 4th Divisions, STRAC consisted of a division's worth of infantry, armor, and cavalry assets arrayed at Forts Devens, Hood, and Meade, as well as the 1st Logistical Command at Fort Bragg. On paper, these units allowed for a self-sufficient force tailorable to any perceived contingency. STRAC was also intended to deter aggression, "so we won't have another Korea," or to prevent, through swift military intervention, escalation to general (nuclear) war. Sink emphasized the need for the predesignation of air force aircraft and a unified commander to ensure readiness. Only then would STRAC become "the force that fills the void between defeat by default and massive retaliation."[32]

Units of the Strategic Army Corps attempted to demonstrate their abilities in training exercises worldwide. These included exercises throughout the Caribbean, in Alaska, the continental United States, Turkey, and even Greenland. From January 14 to February 19, 1959, the 2nd Airborne Battle Group, 503rd Infantry, from the 82nd Airborne Division, flew from Pope Air Force Base, North Carolina, to Alaska to participate in Exercise Caribou Creek. The battle group conducted an air-landed reinforcement in Alaska, followed by tactical parachute operations. As that exercise finished, more than fourteen hundred paratroopers of the 2nd Airborne Battle Group, 501st Infantry—also from the 82nd Airborne Division—flew two thousand miles nonstop in seven hours to Panama in twenty-three C-130 Hercules aircraft. After the long flight, on the morning of February 19, the battle group conducted a parachute-assault exercise to "seize" Rio Hato airfield and then completed a further week of jungle training for Exercise Banyan Tree. Despite their imperfections, both operations successfully showcased

STRAC's mobile capabilities. These missions tested and validated the Strategic Army Corps' ability to operate in disparate climates far from home, a key responsibility of STRAC.[33]

Despite the plethora of exercises, the army's deputy chief of staff for operations, Lt. Gen. John C. Oakes, thought more were needed. In a draft memorandum sent for chief of staff approval, his office wrote that "the credibility of the Army's role as strategic deterrent force is weakened by the lack of frequently demonstrated performances." In his draft proposal for strategic mobility exercises, he envisioned sending reinforced battle group STRAC task forces to places as varied as Turkey, the Philippines, and Thailand throughout fiscal year 1962. These exercises would demonstrate "the strategic capabilities of Army forces," improve "the combat readiness of the Army," and provide "positive evidence to other countries of our ability and determination to participate in the defense of the free world." Oakes reiterated the need to demonstrate a credible capability for deterrence by exercising American expeditionary forces.[34]

Aircraft ranges increased so much by the decade's end that the rapidly deployable Strategic Army Corps was now a national strategic asset on par with the air force's Strategic Air Command. Despite the advances in aircraft, air force support came from two sources—the Tactical Air Command (TAC) and the Military Air Transport Service (MATS). Often, TAC aircraft were disapproved for strategic mobility exercises. At the same time, the army often competed with the air force's other various requirements for MATS aircraft, receiving no more than 35 percent of available flying hours for fiscal year 1959. Nevertheless, while the air force procured improved aircraft, the army developed lighter, air-droppable equipment. Lighter metals allowed the airborne division to maintain the same structure as a regular infantry division. In World War II, pack howitzers, mortars, and bazookas were the heaviest weapons airborne troops could hope to bring into combat by parachute. By 1960, artillery and vehicles were not only air transportable but air droppable.[35]

Strike Command

Strategic response capabilities expanded in the early 1960s. The XVIII Airborne Corps commander Lt. Gen. Thomas Trapnell, who endured the Bataan Death March and later commanded the 187th Airborne Regimental Combat Team in Korea, worried that the navy and marines were attempting to infringe on the army and air force's natural role and mission for rapid strategic response. He recommended a joint "briefing to showcase how, together, those elements can engage and defeat any aggressor in any area of the world, without regard to distances, obstacles, or climate." The disparate commands of the 1950s-era response forces needed a new, unified command structure.[36]

On January 1, 1961, the Department of Defense activated United States Strike Command (STRICOM) at MacDill Air Force Base, Florida, under the command of Gen. Paul D. Adams. STRICOM was born of a desire to combine the Strategic Army Corps of US Continental Army Command and the US Air Force TAC into an integrated command structure of ready forces based in the continental United States. This was done partly to alleviate competing demands on aircraft and provide unity of command for any operation. STRICOM's biggest "almost" operation was Operations Plan 316 (OPLAN 316), the proposed invasion of Cuba during the Cuban Missile Crisis of October 1962. By 1965, the Atlantic Fleet joined the command to give the United States a "worldwide General-Purpose Forces Command." STRICOM was a joint command responsible for the planning and executing of multiple exercises and operations from its inception until it was reorganized as the US Readiness Command in 1972. The army's portion was ARSTRIKE, or Army Strike, while the air force provided elements known as AF-STRIKE. The command's initial missions included providing a reserve of general purpose forces for reinforcing the other unified commands, training the general reserve, developing joint doctrine, planning for and executing contingency operations, and providing troops for later civil disturbance operations. The army also expanded its airborne capabilities to provide a "more flexible response" to regional crises, with the activation of the 173rd Airborne Brigade (Separate) in Okinawa and the 1st Brigade (Airborne/Mechanized) of the 8th Infantry Division on March 26 and 27, 1963 respectively. This unique brigade was to be airborne *and* mechanized—an interesting concept attempting to marry two cultures that ultimately proved unsuccessful.[37]

Life in airborne units under STRICOM continued much like Galvin described in the late 1950s. Men and equipment stayed on alert, bags packed, and personal goods ready for storage in a constant rotation of the Division Ready Force and its constituent Immediate Response Force. Units on alert were to be ready to move within eighteen hours of notification. The two-hour recall status was shared around the division, often lasting two weeks for one company before rotating to the next. Galvin also described how his unit packed their gear and slept in pup tents beside the C-124 Globemasters standing by on the taxiway. Galvin's experiences were normalized as part of a unit with an expeditionary mindset. The Division Ready Force consisted of one thousand men within an infantry battalion task force, including a battery of 105mm howitzers and a platoon of engineers. Sometimes, if more advanced notice were given, the ready brigade would move to a secure location and be able to depart even before eighteen hours. Those at the highest state of readiness were on a two-hour recall notice and maintained one company as the Immediate Response Force restricted to the barracks. Once the unit received an alert, the battalion had to be on base, assembled within two hours, and prepared to leave within eighteen hours of the original notification. The nine infantry battalions in the 82nd Airborne Division would rotate through training and support functions throughout the year.[38]

In October 1962, the 82nd and 101st Airborne Divisions went on full alert to prepare to execute a parachute assault into Cuba during the missile crisis. While the XVIII Airborne Corps and the 101st and 82nd Airborne Divisions prepared to seize four airfields, the US Marine Corps was prepared to make an amphibious assault. They were to secure the port of Mariel to bring in Task Force Charlie, an armored brigade from the 1st Armored Division. The rest of the 1st Armored and the 2nd Infantry Divisions would constitute a floating reserve. Once all marine and army assets were on the ground, Hamilton Howze, the commanding general of XVIII Airborne Corps, would assume command of all forces, known as Joint Task Force Cuba. Maj. Gen. William P. Yarborough would dispatch Army Special Forces teams and Cuban exiles to support an uprising under his authority as commander of the Joint Unconventional Warfare Task Force, Atlantic. The commander of STRICOM during the Cuban Missile Crisis was Gen. Paul Adams, who had successfully commanded joint forces during the 1958 Lebanon intervention. Airborne officers commanded most of the major units involved in the Cuban Missile Crisis. Besides the airborne division commanders, Taylor was serving as the chairman of the Joint Chiefs of Staff, Adams commanded STRICOM, Howze led the XVIII Airborne Corps, Yarborough oversaw Special Forces, and Maj. Gen. Charles Billingslea commanded the 2nd Infantry Division. The planned joint seaborne and airborne invasion never happened, of course.[39]

To further exercise the capability of STRAC and MATS to move forces rapidly to reinforce Europe, the 2nd Armored Division executed

Exercise Big Lift in October 1963. Given seventy-two hours to accomplish the feat, the combined team completed the last flights into France and Germany in sixty-four hours. US Air Force transport aircraft had transported 15,358 personnel and 504 tons of equipment in what was, according to Secretary of Defense McNamara, "the transoceanic largest Army–Air Force deployment ever to be made by air." Rather than ship their armored vehicles, the men of the 2nd Armored flew directly to pre-positioned stocks of tanks and armored personnel carriers in Germany, which the army had decided to store there following the 1961 Berlin Crisis. The exercise was designed to demonstrate the United States' ability to project force for rapid and large-scale reinforcement of NATO by air alone. Some NATO allies, however, expressed skepticism in private that while it was a useful peacetime exercise, the concept had little utility in the event of open hostilities with the missile-armed Soviet Union.[40]

After alerting and standing down for Cuba, STRAC and the 82nd Airborne executed a similar Caribbean response mission in 1965. On Saturday, April 24, 1965, a revolution broke out in Santo Domingo, the Dominican capital, and the crisis deepened over the weekend. On April 27, President Lyndon Johnson sent five hundred marines to protect American lives and property in the Dominican Republic and begin an evacuation in a mission dubbed Operation Power Pack. Two days later, the 82nd Airborne Division reinforced the US Marines already there. At 4:30 p.m. on April 29, the Joint Chiefs of Staff designated Maj. Gen. Robert York of the 82nd Airborne Division commander of all US ground forces in the Dominican Republic and ordered him to deploy the Division Ready Brigade (3rd Brigade with two parachute infantry battalions) to Ramey Air Force Base in Puerto Rico. The original plan was for the 150 C-130 Hercules aircraft to land in Puerto Rico, the men to put on parachutes, and then jump into the capital of Santo Domingo. However, while the force was en route, tensions escalated, and the troopers were needed immediately. The paratroopers received orders to land directly at San Isidro Airfield east of the capital—then in friendly hands. The first elements landed at San Isidro at about 2:15 a.m. on April 30, 1965, in the division's first overseas combat operation since World War II. More elements left Fort Bragg; less than seventy-two hours after notification, two battalions and the division headquarters were on the ground.[41]

Lt. Gen. Bruce Palmer, a Maxwell Taylor protégé, and his XVIII Airborne Corps assumed control of the entire operation, while York directed the ground combat troops. The American mission was to protect the lives of Americans and others and to "give the InterAmerican System a chance to deal with the situation," to preserve law and order, and to prevent a Communist takeover. A multinational force from the Organization of American States joined the Americans, including troops from Brazil, Costa Rica, El Salvador, Honduras, Nicaragua, and Paraguay. Palmer's unstated but obvious mission was to "prevent another Cuba." The initial aim of the 3rd Brigade of the 82nd Airborne Division was to create a safe corridor for US troops through the capital city of Santo Domingo. After securing the primary east–west transportation route through the capital and nicknaming it the "All-American Expressway," York "marched the division band all the way through the corridor." (Fort Liberty, the home of the 82nd Airborne, has a highway, completed in 1978 and dubbed the All-American Freeway, connecting the center of the base with the Cape Fear Valley Medical Center.) Follow-on operations resulted in only small engagements. The pro-Cuban rebels were suppressed, and the legal government reestablished its authority.[42]

Raymond Weaver, a veteran of the 82nd Airborne, recalled, "There was pretty much chaos. And there had been civil but—all kinds of civil strife and lots of people killed down there, lots of civilians killed." As units prepared to head to the island, rumors of fighting grew. Weaver remembered, "We found out that—we heard before we left that the 505[th Parachute Infantry Regiment] . . . had been fired on as they landed." Combat, however, was minimal. This miffed many members of the 82nd, who dreamed of a combat jump and were ill-prepared for governance tasks, yet succeeded anyway. The fighting occurred almost exclusively in an urban environment, facing swift-moving guerrilla forces unlike anything the force had faced before. One soldier remarked, "We would hand out food to the people one minute and then be engaged in a firefight with the same ones the next," wondering why his unit was cleaning the streets since, "Hell, we came here to fight." The entire division was in the Dominican Republic by the end of May. While helping make food, water, and medical care available to the inhabitants, the division also found time to resume proficiency jumps. An eventual diplomatic solution ended the civil war, and most of the division returned to Fort Bragg in June and July, while the division's 1st Brigade remained as a peacekeeping force until September 1966. By the end of their deployment, roughly twelve thousand members of the 82nd Airborne Division had served in the Dominican Republic. But it was not without fighting—27 Americans were killed in action and 172 wounded, the majority from the 82nd Airborne Division. The paratroopers played an instrumental role in ending the civil war, especially thanks to their ability to adapt to the changing political situation on the island. The 82nd Airborne Division demonstrated that regular army units can conduct governance tasks while validating the rapid deployment concept.[43]

Meanwhile in 1965, US commanders in Vietnam had requested combat battalions as the military situation continued to deteriorate. The first army unit that deployed, in April 1965, was the 173rd Airborne Brigade from Okinawa, which was sent to reinforce and defend Bien Hoa Airfield. The plan was for a brigade of the 101st Airborne Division to relieve the 173rd, but that scheme never materialized. Rumors of deployment abounded at Fort Bragg, however. Weaver, the 82nd Airborne veteran of the Dominican intervention, remembered, "The word was the 101st was going one place, and we were going to the other, but we didn't know for sure which was which." As the 82nd prepared for its mission in the Caribbean, the lead elements of the 101st Airborne Division—its 1st Brigade—went to Vietnam, arriving on July 29, 1965. By 1966, personnel from the 101st and the 82nd were sending their men to fill critical personnel shortages for units overseas. The rest of the 101st Airborne Division deployed to Vietnam in 1967. The 173rd Airborne Brigade spent five and a half years "in country" and was employed extensively as a theater-level fire brigade, harnessing an airborne brigade's light, deployable nature. The 173rd even conducted the only large-scale parachute jump of the war during Operation Junction City in 1967.[44]

Later, after the Tet Offensive in early 1968, American leadership requested reinforcements, and the army ordered the 3rd Brigade, 82nd Airborne Division, to move to Vietnam. Desperate for additional forces, Gen. Westmoreland asked for the entire 82nd Airborne Division. The division was a key strategic reaction force and one of the few ready units in the continental United States earmarked for potential conflict in Europe or other "brushfires," and consequently only its 3rd "Golden" Brigade went to Vietnam—six months after it participated in operations controlling civilian rioters in Detroit. The brigade was short-staffed and drew personnel and equipment from 1st and 2nd Brigades to bring it to 95 percent strength. Some 80 percent of the brigade had prior experience in South Vietnam. Within twenty-four hours of receipt of the mission, the initial elements of the brigade were en route, landing on Valentine's Day, 1968. After an airlift of 135 C-141 Starlifter and 6 C-133 Cargomaster aircraft, the brigade arrived at its camp at Chu Lai.

During its tour, the brigade demonstrated immense flexibility, moving between parent commands and locations throughout South Vietnam. The brigade stayed "in country" for approximately a year, departing on February 11, 1969. While the 3rd Brigade served in Vietnam, the army created a 4th Brigade at Fort Bragg to maintain three ready brigades to respond to crises at home or abroad.[45]

Increasing domestic strife during this era gave the airborne another mission—augmenting civilian police. Much as they had in the 1957 Little Rock deployment, rapid-response units functioned as a domestic "brushfire" force as much as an international one. Their quick-reaction capability had even made the 82nd an enticing option for President Truman during a seventeen-day strike by the United Mine Workers in 1946. Labor unrest, however, represented a small fraction of the deployments of troops after World War II. The increasing strife of the civil rights era provided ample opportunity for national leaders to deploy the army's quick-response forces. Besides the movement to Little Rock, airborne units were sent to Oxford, Mississippi, in 1962, to Detroit in 1967, and to Washington, DC, in 1968.[46]

The initial deployment of troops to Oxford, Mississippi, in September 1962 for Operation Rapid Road saw army leadership hesitant to use airborne troops because of heightened Cold War tensions and the potential need for a quick-response force elsewhere. Nevertheless, they were the ideal quick-response force to assist in preventing white violence against James Meredith's attempt to attend classes at and desegregate the University of Mississippi. Airborne soldiers were far from the only military forces that responded to the rioting. By the zenith of tension in Oxford, ten airborne battle groups and command elements of the XVIII Airborne Corps were on the ground at Columbus Air Force Base southeast of the city, under the command of Lt. Gen. Howze, who suggested to the secretary of the army that they execute a parachute drop over Oxford as "a useful training exercise." This rapid deployment happened during planning for the Cuban Missile Crisis, demonstrating the flexibility inherent in the Strategic Army Corps.[47]

In Michigan, after the outbreak of racial violence in Detroit following a police raid, the 82nd and 101st Airborne Divisions each deployed a brigade on July 24, 1967, under the command of Maj. Gen. Richard J. Seitz. Each brigade was ready to move within two hours of notification from XVIII Airborne Corps headquarters and had its lead elements on the ground within ten hours. The paratroopers from both divisions operated out of the state fairgrounds and local schools, with city buses made available for transport. On their second night in the city, the task force (dubbed Task Force Detroit) reported eight "sniper incidents" but nothing more serious than that. Tensions de-escalated, and on the morning of July 26, Lt. Gen. John L. Throckmorton, the XVIII Airborne Corps commander and Seitz's immediate superior, directed all troops to unload weapons and sheath their bayonets. Throckmorton had earlier commanded the 82nd Airborne in a federal response role in Washington, DC, in 1963 and was considered by President Johnson to be ideal for the delicate mission in Michigan. In Detroit, the regular paratroopers proved much more disciplined than local National Guard members, as some 30–40 percent were veterans of the war in Vietnam. In the afternoon, Throckmorton further advised that the paratroopers were to "remove flak vests and do everything to present a return-to-normal appearance to the public." The paratroopers turned over operational control of the area to the 46th Infantry Division of the Michigan

FIGURE 8. An interracial group of army paratroopers take a lunch break while in Detroit for duty during the 1967 riots. Photo from Matthew D. Lassiter and the Policing and Social Justice History-Lab, "Detroit under Fire: Police Violence, Crime Politics, and the Struggle for Racial Justice in the Civil Rights Era" (University of Michigan Carceral State Project, 2021), https://policing.umhistory labs.lsa.umich.edu/s/detroitunderfire/item/4455. Originally compiled by the Public Information Office of the Michigan National Guard.

Army National Guard and redeployed to Fort Campbell and Fort Bragg on August 1 and 2.[48]

More troops were deployed all over the country from various installations, including Fort Bragg, during the turmoil of 1967 and 1968, with the army setting up various task force headquarters in major American cities in operations as well planned as any during World War II. The deployment to Washington, DC, on April 5 and 6, 1968, in the wake of the assassination of Martin Luther King Jr. was a particularly delicate assignment. Paratroopers from the 82nd Airborne Division formed yet another task force, this time under the command of Task Force Washington. The 82nd guarded national landmarks, businesses, and the Capitol building while patrolling the district's east side. The troopers returned to Fort Bragg on April 12. While these well-honed units were well prepared for overseas contingency missions, domestic riot control was often more common. This represents the dichotomy of rapid-response units—these quick-reaction forces were often the only ones available and flexible enough to perform such a mission. The increased discipline expected of elite infantry paid dividends in preventing large-scale incidents between troopers and civilians. Tasked with restoring order alongside the National Guard and regular law-enforcement units, these deployments of federal troops for domestic situations have become an enduring and uncomfortable feature of American life.[49]

Between World War II and Vietnam, airpower drove military innovation, and the army responded through the leadership of airborne officers to create a strategically mobile and air-transportable force responsive to what its leaders envisioned as the future of warfare. Despite massive budget cuts and, at times, contentious relationships with the Joint Chiefs of Staff, the army persevered. But to do so it had to show its worth in a contemporary operating environment that measured response and deployment in hours instead of weeks. In addition to the instances described, small contingents of the 82nd were also deployed to the Congo in 1964 and 1967 to provide security for US Air Force aircraft assisting Belgian forces. The Strategic Army Corps serves as a sign of the long-lasting influence the airborne has had on postwar army strategic capabilities. The rapid-response force developed during the Carter administration and its modern incarnation—the Immediate Response Force—are direct descendants of the Strategic Army Corps developed during Taylor's tenure as army chief of staff.[50]

Airborne leaders' outsize influence in the postwar period put them at the forefront of organizational change. The airborne mafia shaped tactical thought and forced the army to redefine its role during the air-minded atomic age by emphasizing strategic mobility and the quick delivery of land combat power to the battlefield by air. Rapid airborne response capabilities became a critical component of postwar planning. The value of an airborne force, according to Gen. Howze, "lies in its contingency missions, its readiness to move by aircraft and to parachute onto foreign soil, if necessary, almost anywhere in the world that an airplane can get to . . . and thus serve as a very prompt application of force in situations other than that of a major war." The airborne also represented a deterrence component, much akin to the Strategic Air Command of the air force. As Gavin wrote in 1947, "The knowledge of the existence of a well-trained airborne army capable of moving anywhere on the globe on short notice, available to an international security body such as the United Nations, is our best guarantee of lasting peace. And the group of nations that control the air will control the peace." Strategic air mobility would bridge the gaps between the army's personnel reduction, the national focus on airpower, and the military mission of conducting land combat.[51]

In the early Cold War, the ability to move rapidly and resupply forces by air became vital to the nation's quick-response capabilities, whether to reinforce Europe or put out a brushfire elsewhere. The airborne mafia's expeditionary mindset rooted in air-mindedness was critical to the army remaining a viable component of the national security apparatus. Creating a highly mobile ground force capable of worldwide deployment was necessary in an era where air power enthusiasts dominated military thinking, bent on making strategic bombing the only viable way of war. Still, while some have dismissed this capability as only a "paper army" or inconsequential, STRAC, the Strategic Army Corps— alongside missile and helicopter forces—represented army efforts to remain a viable strategic option in a rapidly shifting international environment. Experiences in World War II fueled the development of more and better ways to project force from within the United States to hot spots around the globe, which gave the army a critical component of the country's national defense options. In the army's view, airborne forces were fundamental as an alternative to massive retaliation. The development of strategic response forces in the 1950s set the stage for future applications of expeditionary warfare and has had lasting effects on how the United States projects combat power around the world today.[52]

Epilogue
The Legacy of the Airborne Mafia

> There is no rank in the LGOP. In this division,
> leaders jump first, eat last—always.
>
> —Maj. Gen. Christopher Donahue, 2021

Airborne culture has remained a significant part of the US Army since its inception. The airborne mafia started as a small cadre of radical thinkers who insisted that the fusion of air and land power was critical to future warfare. Since 1940, airborne units have imparted their values, beliefs, and norms to the rest of the army despite decreasing use of parachute-delivered troops in high-intensity combat operations. The cultural tenets identified throughout this work, those of exceptionalism, flexibility, adaptability, innovation, inspired leadership, decentralization, and individuality, have permeated the army and allowed airborne leaders to institutionalize their specialty as a critical component of the army despite a lessening requirement for large-scale airborne operations. Soldiers from all branches and military occupational specialties sing running cadences about C-130s rolling down an airstrip, sending their troopers on a top-secret mission, destination unknown, where they are unsure if they are ever coming home. The values, beliefs, and norms developed in World War II continue to impact the army. Airborne units also continue to play a critical role as rapid-response forces.[1]

The airborne has always been more than a means of delivery to the battlefield, and its importance in the larger army outweighed mere tactical concerns. The very nature of airborne operations required personnel to be mentally prepared to face danger before facing the enemy—jumping from an aircraft in flight carried a high level of uncertainty. As a result, the image of the paratrooper became synonymous with the army's elite, as it has in many other militaries around the world. Moreover, these ideas permeated the rest of the army. By the late 1950s, all new officers were required to pass airborne or ranger school before going to their first units. Airborne training was not required because of its military utility but because leaders believed it instilled discipline and courage in those who partook of it.[2]

The airborne mafia have left a legacy in the army in tactics, strategy, and culture that continues into the twenty-first century. Tactically, the airborne mafia helped the entire army become more air-minded through helicopter assault while also helping to pioneer special operations forces. Strategically, the army maintains a rapid-response capability, while many lessons from the Pentagon tenures of Generals Ridgway, Taylor, and Gavin continue to ring true today. Culturally, the army promotes airborne leaders and emphasizes, as an informal and unwritten prerequisite for promotion, "punching the airborne ticket" by attending either airborne or air assault school in one's career. Those markers of the airborne mafia's legacy are important cultural artifacts in today's army. Likewise, airborne units hark back to their World War II legacy and cultural tenets to build a strong esprit. Airborne culture is a significant part of army culture.

Tactics

Thanks to the airborne mafia, the army remains "air-minded," as it has developed techniques and ideas since the 1960s that have kept the service agile, adaptable, and innovative. This air-mindedness has resulted in tactical innovations rooted in the original conception of how airborne units would operate and in the reality of World War II airborne operations. The modern air assault concept, special operations forces, and doctrinal changes are important tactical legacies of the airborne mafia found in the twenty-first-century US Army. Even the men who led American forces in the Vietnam War were, by and large, paratroopers.

Through the Vietnam War, airborne officer protégés of Ridgway, Taylor, and Gavin were in the forefront of the American effort. Despite his consistent early warnings of a widening commitment, Maxwell Taylor was a key player in American involvement in Vietnam while serving as chairman of the Joint Chiefs and later ambassador to South Vietnam.

Meanwhile, his protégé William Westmoreland led Military Assistance Command, Vietnam (MACV) from 1964 to 1968. Other officers with World War II experience in airborne units commanded divisions and corps throughout the war, including Melvin Zais, Julian Ewell, John Norton, and John Tolson, who served, respectively, in the 517th, 501st, 505th, and 503rd Parachute Infantry Regiments during World War II. Richard Seitz, another 517th veteran, served in numerous key staff positions for Westmoreland and commanded the 82nd Airborne Division when it sent a brigade to Southeast Asia and troops to quell domestic civil unrest in 1967 and 1968. Five airborne brigades were deployed to Vietnam at one time or another, including the 101st Airborne Division, the 173rd Airborne Brigade, and the 3rd Brigade, 82nd Airborne Division.[3]

Helicopter air mobility continued to evolve in the early part of the war. The People's Army of Vietnam (PAVN) and the People's Liberation Armed Forces (Vietcong) learned to negate the advantage of air mobility by not engaging large American forces and staying close to the Americans in the jungle to offset superior firepower. By the end of 1966, the 1st Cavalry Division, for example, focused on foot patrols and ambushes, and half its contacts came from a third of its forces—those

FIGURE 9.   Capt. Thomas Taylor, *left*, of the 1st Brigade, 101st Airborne Division, arrives in Vietnam in July 1965, on the same day that his father, Gen. Maxwell Taylor, left Vietnam. US Army photo.

operating on foot. Nevertheless, the utility of helicopters was so evident that the army decided to divide the rotary-wing assets in-country among all combat units. This was done so that every brigade had an organic assault helicopter capability, although this was not enough to move more than a company of infantry at once. Still, this reflects the ubiquity of the airmobile concept and the legacy of the airborne mafia.[4]

Aside from the "Triple Capability" TRICAP division, air mobility fell out of favor in the 1970s. In the lingering aftermath of the American defeat in Vietnam, leaders questioned the survivability and usefulness of helicopters on the new mechanized battlefield as envisioned in the 1976 Field Manual (FM) 100-5. The doctrine of the day, commonly referred to as "active defense," assumed numerical inferiority and was predicated on massing firepower in tank combat. Rather than focus on light forces moving about the battlefield by air, leaders were more interested in employing helicopters for antitank warfare, epitomized by the development of the AH-64 Apache attack helicopter. By the 1980s, helicopters were again in vogue for missions other than antitank roles. They became more significant to army planning thanks to the introduction of "AirLand Battle" doctrine in the 1982 version of FM 100-5.[5]

The 101st Airborne Division is still the only full-scale air assault division in the army—though every division has a combat aviation brigade and the ability to lift at least a portion of its combat personnel at any given time. What distinguishes the 101st Airborne from other army divisions is its institutional focus and expertise in air assault operations. The rest of the army considers the Screaming Eagles the lead proponent of modern air assault tactics, techniques, and procedures. As of 2024, the army has twelve combat aviation brigades, including eleven divisional and one separate in the active component, with eight divisional and six separate in the US Army Reserve and National Guard. The March 25, 2003, deep penetration raid in the opening stages of Operation Iraqi Freedom by thirty AH-64 Apaches of the 11th Attack Helicopter Regiment resulted in nearly every aircraft sustaining battle damage, two downed aircraft, and two pilots captured. The mission further highlighted the antiarmor capabilities Hamilton Howze alluded to in his prepared briefings around the Pentagon in the mid-1950s. However, the theorists miscalculated the extent of air defenses on the modern battlefield, making independent deep operations by attack aviation units increasingly difficult.[6]

Throughout subsequent counterinsurgency wars in Iraq and Afghanistan, attack and reconnaissance helicopters served in important cavalry-type roles, assisting ground forces in finding and fixing insurgents as intended but rarely operating on their own without ground force assistance. UH-60 Blackhawks and venerable CH-47 Chinooks also served prominent roles in moving troops into and out of combat, especially in the rugged mountainous terrain of Afghanistan. Furthermore, helicopter medical evacuation has played an important role—evolving from its origins in World War II through difficult lessons in Vietnam—in decreasing mortality by reducing transport time. This capability is a direct result of the efforts of Ridgway, Taylor, and Gavin in World War II and, more proximately, the evolution of the airborne concept into airmobile techniques.[7]

The tactical impact of airborne leaders is best summarized by examining terms in the army's capstone doctrine for operations. I analyzed the frequency of the words "airborne," "airmobile," and "air assault" against the terms "armor(ed)" and "tank(s)" across each of the twelve different publications of FM 100–5 or FM 3–0 published between 1949 and 2022. The term "airborne" peaks in the 1968 version, alongside "airmobile," before being superseded by tanks and armored warfare in 1976. The words have stabilized since 1976, yet armor-related terms remain slightly more prevalent today. Those numbers indicate an institutional tactical preference for aerial-type operations as the airborne mafia was in charge through the Vietnam War. In many ways, the war in Vietnam was the airborne mafia's war. The preponderance of units that fought there were light infantry—initially mostly airborne and Special Forces. Tactically, the war was dominated by vertical envelopment through air mobility with helicopters. Portions of all major airborne units took part. The service's attempts to learn counterinsurgency during the 1950s and early 1960s were championed by many airborne officers, and the overall commander in Vietnam, Gen. William C. Westmoreland, was himself a member of the airborne mafia. US Army doctrinal data also demonstrates a clear shift to focus on mechanization in the 1976 manual, which boasted that one of every two American infantrymen was fully mechanized.[8]

Overall, this suggests a distinct post-Vietnam cultural shift in the army, especially regarding tactics. Supplanting the airborne, armor officers and mechanized doctrine came to dominate the service. Instead of maintaining a flexible army prepared for multiple threat scenarios across the spectrum of warfare, the US Army of the late 1970s and 1980s focused almost exclusively on large-scale armored combat. The 1973 Yom Kippur War helped thrust the army in that direction as William

E. Depuy (a onetime Taylor acolyte) and others rewrote army doctrine with a nearly singular focus on large-scale mechanized forces to meet the threat of the Soviet Red Army in Central Europe. This came at the expense of learning from the American experience in Vietnam, alongside an attempted institutional erasure of counterinsurgency ideas. One only needs to compare the 1962 and 1976 editions of FM 100–5 to understand the vast changes in doctrine and cultural emphasis. Gavin was consulted to assist on the 1976 manual but called it a manual "that in some places appeared a compendium of opinions." Nevertheless, an old airborne officer, Lt. Gen. John Norton, commanded Combat Developments Command between 1970 and 1973, pivotal years during which it developed the requirements that led to the "big five" weapons systems developed during the 1980s. These included the M-1 Abrams main battle tank, the M-2 Bradley infantry fighting vehicle, the UH-60 Black Hawk transport helicopter, the AH-64 Apache attack helicopter, and the MIM-104 Patriot air defense missile. While the helicopters were heavier and less air transportable across long distances, they reflected a continued commitment to the fusion of air and ground forces to defeat Soviet armor. Even airborne officers had a hand in the mechanization of the post-Vietnam army.[9]

Strategy

The airborne mafia have also had a lasting effect on civil-military relations and American strategic culture, in the organization of the Joint Chiefs of Staff and their relationship with the president, and the evolution of parachute-capable rapid-response forces. The airborne mafia were correct in their critique of massive retaliation, as increased tensions in disparate parts of the world, including Southeast Asia and the Middle East, caused President Eisenhower to quietly reconsider his stance on limited warfare and the utility of conventional forces. However, the airborne mafia's legacy in civil-military relations furthered the gap between the military sphere and political spheres of power in Washington. The 1958 Defense Reorganization Act reduced some of the service chiefs' responsibilities, removed them from the chain of command in wartime, and emphasized the role of the chairman. By the time President Kennedy took office, the structure of the Joint Chiefs actively discouraged broad dissenting advice, which separated the president from the service chiefs and prevented him from receiving differing opinions while making critical strategic decisions.[10] All three airborne officers were saying the same thing in different ways. They all held that nuclear warfare was so unimaginable that warfare would revert to its natural form—ground forces fighting for territorial domination. They argued for well-trained and well-resourced forces to fight a ground war in various forms in any theater of the world.

The airborne mafia also had a tangible and lasting impact on strategic force structure. The continuing emphasis on quick-response forces to give national leaders a flexible parachute-capable option reflects the expeditionary mindset developed from the 1940s through the 1960s. While many airborne officers and units fought in Southeast Asia, the XVIII Airborne Corps and most of the 82nd Airborne Division were held in reserve to fulfill the STRAC mission of rapid response. They were used throughout the continental United States but remained— on paper—capable of worldwide deployment to meet contingencies. Of course, with the constant rotation of military personnel into and out of Vietnam, the 82nd Airborne Division was always in transition and often understrength, despite its priority status. Nevertheless, the division was placed on alert three times throughout the 1970s. First, in 1973, in case the Yom Kippur War might have expanded; second, in 1978, to respond to tensions in Zaire; and third, for the 1979 Iran hostage crisis.[11]

Meanwhile, the United States increased its quick-response-force capabilities. In 1974, the first two modern Ranger battalions were formed on the orders of army chief of staff Gen. Creighton Abrams (an armor officer for most of his career). The army reactivated the 1st and 2nd Ranger Battalions as elite, light, highly proficient infantry battalions. Since its inception, the 75th Ranger Regiment has maintained a special operations role, prepared to conduct no-notice forced-entry operations, deep penetration raids, and provide support to other special operations units, while it carries on the legacy of airborne elitism today. Its leaders are promoted at a higher rate than other infantry officers, and the unit is charged with disseminating its knowledge and expertise to the rest of the force, effectively diffusing a new "Ranger culture" that is an evolved version of airborne culture.[12]

After the Soviet invasion of Afghanistan and during the Iran hostage crisis, the Carter administration pushed forward a new Rapid Deployment Joint Task Force (RDJTF) headquarters that would command forces assigned to contingency operations. Activated on March 1, 1980, the RDJTF was tasked with out-loading and supporting American quick-reaction forces and figuring out what equipment was needed to move forces quickly. The idea was to tailor equipment to the force rather

than the force to available means of transport. The RDJTF was initially designed as a light, general purpose force to respond with special
quickness to contingencies worldwide. However, given existing plans
for defense against a conjectural twenty-two-division Soviet invasion of
Iran, the task force evolved into a large, heavy, and firepower-dependent
group. Nevertheless, the RDJTF was intended to fight any number of
"half wars," small wars, or contingency operations in those areas. At the
same time, the Department of Defense maintained enough resources to
fight the Soviets in Europe.[13]

Refining concepts like rapid-response forces earmarked for opening
airheads to support contingency operations worldwide continued into
Ronald Reagan's presidency. In his 1981 article in the *New York Times*
that elaborated on new ideas in the defense strategy of the Reagan administration, incoming CIA director Stansfield Turner extolled the virtues of rapid-response forces as crucial to combat threats in the Persian
Gulf. Gambling that decisive rapid movement of American forces to
seize bases would deter Soviet interference in the region, Turner cited
the inability of the United States to quickly move sufficient forces into
Korea in 1950 as proof of the need to maintain robust quick-response-
force capability. Congress offered solutions to recapture the "rapid"
portion of the RDJTF, namely to assign the XVIII Airborne Corps as
the permanent army component with its two airborne divisions, the
82nd and 101st—like in Bastogne and the early Cold War.[14]

The American interventions into Grenada and Panama during the
1980s represent activations of rapid-response units that have become
part of airborne lore. When Saddam Hussein invaded Kuwait and
threatened Saudi Arabia, the army initially deployed the 82nd Airborne
Division to the Persian Gulf because of its rapid-response capability.
The first members of the division left the United States on August 8,
1990, and the entire Division Ready Brigade was on the ground in
Saudi Arabia and ready to fight seven days after the alert. By August 24,
the twelve thousand paratroopers from the division were staged and
ready in Saudi Arabia, an incredible feat that took 582 C-141 sorties.
Further, throughout the 1990s, the army focused on *early* entry forces
designed to follow an initial forced-entry assault, fight, and buy time
for heavier forces. The speed with which light, airborne forces could
deploy was unmatched by heavier, armored formations that sometimes
took weeks to arrive via the fastest ships. The army needed something
in between and developed a new unit type based on a lightly armored
vehicle—the Stryker—that could carry a standard infantry squad but

was C-130 transportable and therefore could be landed on unimproved runways. The only version of the Stryker that is not air transportable is the double V-hull version that was developed specifically in response to improvised explosive device threats encountered by US forces in Iraq in the early 2000s. The Stryker gives the army an "early entry" capability to augment airborne or ranger forcible entry operations.[15]

Since the attacks of September 11, 2001, airborne culture, in all its many forms, has played a critical role in the US Army's prosecution of the wars in Iraq and Afghanistan. Parachute and heliborne forces seized footholds in Afghanistan in 2001 and were used in Iraq in 2003. Small 75th Ranger Regiment parachute operations occurred near Kandahar, Afghanistan, and later in the deserts of western Iraq. These operations provided American planners with airfields to support operations in both countries. The first brigade committed to Iraq for the 2007 troop surge was the 2nd Brigade, 82nd Airborne Division, thanks to its assumption of the Global Response Force mission at the time. Initially deployed to Kuwait in January 2007 as the theater reserve, it moved into Baghdad to support multinational force operations in Iraq to stem sectarian violence over the next fifteen months.[16]

Throughout the twenty-first century, the obligation of rapid response continued to fall to airborne forces. Today, the 82nd Airborne Division has a mission to be prepared to deploy elements of its formation anywhere in the world. Initially, a battalion-size task force can be "wheels up" on its way to an objective in eighteen hours, and within thirty-six hours an entire brigade's forty-three hundred paratroopers can be in the air. The former Global Response Force (GRF) deployed sixteen times to buttress operations in Iraq and Afghanistan between 2001 and 2013 and twice more for humanitarian operations. In 2010, the GRF supported relief operations following an earthquake in Haiti and almost immediately sent a battalion to Afghanistan after its return.[17]

In the 2020s, this force is called the Immediate Response Force, which consists of three subordinate Immediate Response Battalions at varying levels of echeloned readiness. The IRF deployed three times overseas between 2020 and 2022: first in Iraq after Iranian missiles threatened the American base in Baghdad on New Year's Day 2020; in 2021 to supervise the ad hoc departure of American forces from Afghanistan in August; and in 2022 to Europe in response to the Russian invasion of Ukraine. In each case, the 82nd's ready battalion deployed within eighteen hours. One of the significant side effects of fostering an expeditionary mindset is a cultural predisposition toward readiness that manifests itself in the necessary training and resources. As the journalist Jay Price put it, "In one sense, the 82nd's specialty is parachuting. In another, it is being ready." This capability gives the United States unprecedented flexibility in deciding options for national security crises or even natural disasters.[18]

In the summer of 2020, during the Black Lives Matter protests, the 82nd's IRF deployed alongside other forces from throughout the country to the National Capital Region. Much like the airborne forces, the 16th Military Police Brigade headquarters and its 91st Military Police Battalion also deployed rapidly and had their full complement of soldiers in Washington within seventy-two hours. As in the army's response to domestic situations in the 1950s and '60s, the paratroopers took bayonets, which later became a source of media controversy. The bayonets were quickly placed inside duffel bags, and the 2nd Battalion, 504th Infantry, typically called Task Force White Devil, changed its name to TF 2–504. The paratroopers staged at Andrews Air Force Base but never entered the city. This battalion had deployed to Iraq in January 2020 and was the core of the same task force activated for Afghanistan in August 2021.[19]

During the initial confrontation with Vladimir Putin and the Russian Federation over the invasion of Ukraine in early 2022, the 82nd Airborne Division and XVIII Airborne Corps were the first American forces moved to Europe to bolster NATO—as the STRAC was designed to do in the late 1950s. The deployment in Poland was the fourth for the 82nd Airborne Division's IRF in two years. The XVIII Airborne Corps, commanded by Lt. Gen. Michael Erik Kurilla, a former commander of both the 75th Ranger Regiment and the 82nd Airborne Division, set up a headquarters at Wiesbaden, Germany, as Task Force Dragon, while the 82nd Airborne Division moved into Poland. The XVIII Airborne Corps was sent to form the command element of a joint task force designed to integrate units and assets across service and multinational lines should the situation call for it. The decision to send the 82nd Airborne and its corps headquarters is a notable sign of how deeply embedded airborne ideas from the 1950s are woven into the fabric of the army. That the United States would first send lightly equipped paratroopers to stare down a heavily mechanized Russian force speaks volumes about how trusted airborne forces are and how normalized using airborne units in all manner of missions has become. When the country needs a quick-reaction force to fly across the world, it routinely calls parachute-capable forces with either a direct lineage to early airborne units or derivatives thereof. The 82nd, because it maintains forces that are poised to deploy within eighteen hours of notification, has become a national "easy button." It has become, over eighty years, America's all-purpose force.[20]

Culture

Airborne culture is a significant component of the army's overall ethos. The airborne mafia remained alive and well throughout the twentieth century. In 1988, Maj. Gen. Bernard Loeffke congratulated Maj. Gen. John Foss on his appointment as the US Army's deputy chief of staff for operations and noted that "we now have the airborne mafia in charge of DCSOPS." A 1992 Naval War College paper on redefining roles and missions for the United States military referenced the list of officers who had served in the 82nd Airborne Division as a "Who's Who of the 20th Century Army," forming a formidable mafia committed to self-preservation. No other branch, not the infantry, armor, artillery, or even the cavalry, has its own nationally recognized holiday, yet the airborne does. When President George W. Bush declared August 16, 2002, as "National Airborne Day" to commemorate the first official parachute jump by the test platoon, he put the airborne on par with the other services in giving national status to its birthday. Senate Resolution 235 entrenched the day as a holiday in 2009. Later, when the United States Military Academy at West Point created special football uniforms to honor the past, the first choice—in 2016—was the 82nd Airborne Division and a World War II scheme. The 1st Cavalry Division (Airmobile) from Vietnam was selected in 2019.[21]

Nearly all combat arms officers attend airborne or air assault training, an important milestone for junior officers' careers. Of the thirty-one brigade combat teams active in the US Army as of 2022, five are airborne, and another three are air assault brigades in the 101st Airborne Division (Air Assault), while the army reactivated the 11th Airborne Division in 2022 to command units in Alaska. This marks the first time the United States has had three active airborne divisions since 1958. Even one of the legends that explains the origins of the army's ubiquitous "Hooah" is attributed to the 82nd Airborne: this particular version credits members of the division in World War II who responded to their commanders with "HUA," an acronym meaning "heard, understood, acknowledged." Further, sixteen of the twenty-one army chiefs of staff since Ridgway display their parachutist badges in their official photos.[22]

Another international symbol of the airborne, the maroon beret, has been worn by paratroopers since 1981, when Gen. Edward Meyer authorized its use. The saga of berets before that time is very long and contentious; however, the first unit in the US Army to wear the maroon beret was the 509th Parachute Infantry Battalion, upon being gifted the symbol by the British 1st Airborne Corps commander during the Second World War. In 1973, the maroon beret was authorized by the 82nd Airborne division commander. But in 1978, when many unofficial colored berets permeated the force, the army chief of staff Gen. Bernard W. Rogers deauthorized all berets except the green Special Forces headgear, citing the berets' proliferation as detrimental to "Army wide uniformity." A grassroots "National Committee to Save the Beret" emerged, including Ridgway, Gavin, and influential senators. This resulted in a Senate resolution asking the secretary of the army to reverse the ban. Rogers's successor as chief of staff, Gen. Edward C. Meyer—himself a former 82nd Airborne commander—noted that he had received "more advice, guidance, and counsel on the beret than I have had on whether or not we go to war." He reversed the decision for the black Ranger beret in 1979, and in 1981 the maroon beret was authorized for the 82nd Airborne Division. It continues to be integral to the airborne's distinctive appearance today.[23]

Senior and Master Parachutist Badges represent another cultural artifact of airborne units. They were authorized in 1949 and denoted an expert in military static-line parachute operations and are awarded to "jumpmasters" with more than thirty and sixty-five jumps, respectively. Between these badges, the Basic Parachutist Badge, and the Air Assault Badge, soldiers can earn markers of their status as trained in these skills and wear them throughout the army—whether assigned to those type units or not. These cultural artifacts automatically elevate their wearers within whatever non-airborne or non-air-assault type unit they may be assigned to by denoting them as a two-time volunteer who has passed one of those essential rites of passage.[24]

The culture ingrained in airborne units in World War II plays an instrumental role in how those units see themselves and operate to this day. The 82nd Airborne Division War Memorial Museum at Fort Liberty (formerly Fort Bragg) helps maintain an appreciation for the past within the 82nd Airborne Division. While every major army installation has a museum, new paratroopers assigned to the 82nd Airborne are brought to its museum explicitly to inculcate them into the culture and history of the unit of which they are now a member. The visit is "a way of maintaining a consistent culture and set of values in an organization that has constant turnover as soldiers cycle in and out—far more turnover than most civilian companies experience," according to reporter Jay Price. This cultural immersion helps inculcate in every new paratrooper the values, beliefs, and norms that form the core of the division's institutional makeup.[25]

The 82nd Airborne's museum began in a room in Berlin and moved Fort Bragg when the division returned in 1946. When its permanent building opened in 1957, it was the first purpose-built museum in the army. The 101st Airborne Division has its own division museum at Fort Campbell, and the Airborne and Special Operations Museum in downtown Fayetteville, North Carolina, helps cement the legacy of these elite forces throughout history. To further cement the connection between modern paratroopers and the legacy of the Second World War, hundreds of paratroopers from across the airborne community travel to Normandy annually to pay respects to their predecessors. Modern paratroopers continue to cement their connection to the original pioneers by participating in remembrance ceremonies and a parachute jump near Sainte-Mère-Église every year.[26]

On August 12, 2021, the 82nd was sent to assist in the United States' largest noncombatant evacuation operation. The paratroopers from the 82nd took charge of security around the airfield. They were ideally suited for the mission because of their readiness as the IRF and institutional focus on airfield-related missions. While the overall operation was a joint effort, the last soldier to leave Afghanistan at the end of the nearly twenty-year American commitment there was the commander of the 82nd Airborne Division, Maj. Gen. Chris Donahue. His decision to be the last American soldier to leave Afghanistan reflected more than eighty years of ingrained airborne leader culture—first in, last out. His boots are now on display at the new National Museum of the United States Army outside Fort Belvoir, Virginia. "In this division, leaders jump first, eat last—always," Donahue said.[27]

Reflecting the sort of ingenuity in ambiguous situations known to their World War II forebears, American paratroopers traded two cans of chewing tobacco for a green Toyota Land Cruiser pickup truck outfitted with a Russian-made ZPU-2 antiaircraft gun on August 17, 2021. To operate the vehicle, they relied on an Iraqi-born paratrooper, Pfc. Alsajjad al Lami from the 1st Battalion, 504th Parachute Infantry, who had served in the Iraqi military before emigrating to the United States. The truck-mounted gun was bigger than the troopers' primary infantry

FIGURE 10. Maj. Gen. Chris Donahue, commander of the US Army 82nd Airborne Division, XVIII Airborne Corps, boards a C-17 cargo plane at the Hamid Karzai International Airport in Kabul, Afghanistan, on August 30, 2021. US Army photo by Master Sgt. Alexander Burnett, US Central Command Public Affairs.

weapons and served as a deterrent to Taliban fighters as US forces landed in Kabul to assist with the massive evacuation effort there. The troopers had improvised a timely "show of force" to the Taliban surrounding the airport. (The truck itself was returned from Afghanistan and placed at the 82nd Airborne Division War Memorial and Museum.) The 82nd's efforts in Kabul contributed to the evacuation of more than 120,000 Afghans and Western civilians between August 14 and 31, 2021, as part of their noncombatant evacuation operation. At one point, more than half the US Air Force's 222 C-17 Globemaster III aircraft were running around-the-clock missions evacuating people and bringing supplies to Americans on the ground. As bookends to

America's longest war, which began with Special Forces, Ranger, Marine, and parachute units landing in Afghanistan, parachute units were there at the end as well.[28]

To this day, airborne culture continues to shape the US Army. In 2022, the army reactivated the 11th Airborne Division to take over US Army Alaska as part of its long-term Arctic strategy. The army chose the 11th Airborne in lieu of the 6th Infantry Division, which had a longer history in Alaska. This paradox between decreasing operational use (few parachute operations since 1945) and their increasing cultural influence (the choice of the 11th over the 6th) is a clear sign of the long-term impact of the airborne mafia. Airborne paraphernalia adorns most army posts, and seemingly every soldier in the army knows the cadence about C-130s rolling down the strip. Early twenty-first-century films like *Band of Brothers*, *We Were Soldiers*, and *Black Hawk Down* further help inculcate airborne lore. Large-scale airborne operations may be outdated, and airborne units might one day be relegated to the historical dustbin like the old (also once elite) horse cavalry. However, the impact of the airborne mafia in implementing lasting, meaningful change in the army's organizational culture and institutional makeup will endure in the overall air-minded expeditionary ethos of the modern US Army.[29]

Airborne warfare still proves exceptionally costly in terms of both money and potential casualties. Gavin realized the futility of airborne operations in the missile age shortly after World War II, and the record since suggests he was mostly right. The only major *parachute* operations since World War II have been performed against lightly armed adversaries. Having a highly trained force capable of going anywhere in the world within eighteen hours is still a viable use of resources for a country that seeks to influence all corners of the globe with military power. The parachute is an inefficient bonus. Airborne forces remain a crucial component of the modern army not because of successful daring operations but rather owing to the institutionalization of their culture within the entire army. Following World War II, America's airborne forces survived because of their ability to adapt and reorganize themselves to suit whatever mission was paramount within the US national defense strategy. The airborne's elite status attracted some of the best and brightest officers; naturally, these leaders rose to high ranks in the army. This feedback loop created a preponderance of former paratroopers leading the army, which allowed the airborne to stay relevant.[30] The use of paratroopers was not a particularly efficient battlefield innovation, as they suffered massive casualties during the Second World War.

However, their legacy persists in places like Fort Liberty, North Carolina; Fort Richardson, Alaska; Fort Campbell, Kentucky; and Vicenza, Italy. While it is unlikely that any modern paratroopers will experience anything like what their forerunners who jumped into Normandy encountered on June 6, 1944, every modern paratrooper accepts (and even welcomes) this possibility—no matter how far-fetched it might be.

The unique culture engendered in World War II airborne units has manifested itself in other ways—primarily in an air-minded expeditionary ethos found in many corners of the army. Thanks to personnel policies that rotate officers and soldiers between units, airborne and air assault troopers routinely find themselves among mechanized units where they can impart the values, beliefs, and norms learned while jumping out of airplanes. Crucially, they can impart the trust inherent in airborne operations—especially trust in one's equipment and comrades, some of whom the paratrooper might not know but must rely on to survive. According to Brig. Gen. Brian Winski, quoted in the *Army Times* article that spurred this book, a "unique environment of absolute trust and subordinating your safety" exists in airborne units. Retired Gen. Barry McCaffery reiterated the importance of this culture that emerges from young soldiers, who are "proud to be doing something complicated, dangerous, and exciting. It's hard to ignore that. That's what comes out of this: young people who want to serve in that kind of unit." This is well summarized in the "rule of the LGOP"—that "little groups of paratroopers" of all ranks and military occupational specialties will accomplish the mission with little guidance—which originated in Sicily and still infuses airborne units.[31]

There is an inherent danger in deploying troops who are always in a high state of readiness and might be primed to shoot first and ask questions later. Because of their intrinsic readiness and well-honed combat skills, parachute forces have become a double-edged sword. High-readiness troops are seldom the best choice for constabulary functions—yet they are often the only choice. Units primed to jump behind enemy lines and fight desperate actions with high casualties would understandably be disappointed to find themselves as police forces brokering peace between warring factions or trying to root out enemies who are in civilian dress in a population-centric counterinsurgency campaign. The first French forces responding to crises in Indochina and Algeria during the 1950s were paratroopers, who also moved toward helicopter tactics. Their brutal techniques, including torture and disdain for civilians, helped undermine the French effort to hang on to its former colonies. French paratroop officers gained notoriety and political clout that turned them into a nightmare of civil-military relations, resulting in their reassignment throughout the French military in 1961. Often placing highly trained, elite units into tense situations had nefarious results, as happened with the British Parachute Regiment in Derry on Bloody Sunday in 1972 and the Canadian Parachute Regiment in Somalia. The British "paras" remain, but the Canadian unit was disbanded.[32]

Nevertheless, the US Army of the twenty-first century, without the impact of Matthew B. Ridgway, Maxwell D. Taylor, James M. Gavin, and their acolytes, would be unrecognizable. The airborne mafia's mindset helped usher in organizational changes centered on decentralization and mobility, helicopter-borne airmobile tactics, and strategic response forces that continue to provide national leaders with a formidable force that can project combat power around the world. The airborne mafia imbued an air-minded expeditionary mindset that has had a lasting impact on how the army organized itself and fought throughout the second half of the twentieth century and into the twenty-first. By institutionalizing a way of thinking that emphasized air-mindedness and sense of immediate response, the airborne mafia profoundly affected the entire army's cultural, strategic, and tactical makeup. Airborne culture has become army culture.

Notes

Introduction

1. Walter Kretchik, Robert F. Baumann, and John T. Fishel, *Invasion, Intervention, "Intervasion": A Concise History of the US Army in Operation Uphold Democracy* (Fort Leavenworth, KS: CGSC Press, 1998), 52; Sean Goldstein, "Talks Barely Beat Invasion US Intervention in Haiti," *Baltimore Sun*, September 20, 1994; Conrad C. Crane, "Phase IV Operations: Where Wars Are Really Won," *Military Review*, May–June 2005, 19.

2. Rachael Riley, "Commander of 82nd Airborne Division to Deploy to Afghanistan," *Fayetteville (NC) Observer*, August 18, 2021, https://www.fayobserver.com/story/news/2021/08/17/pentagon-82nd-airborne-division-role-afghanistan-maj-gen-christopher-donahue-military-biden-taliban/8165782002/; Jeff Schogel, "Why a 2-Star General Was the Last American Service Member to Leave Afghanistan," Task & Purpose, August 31, 2021.

3. Andrew Bacevich, *The Pentomic Era: The US Army between Korea and Vietnam* (Washington, DC: National Defense University, 1986), 107; Walter E. Kretchik, *U.S. Army Doctrine: From the American Revolution to the War on Terror* (Lawrence: University Press of Kansas, 2011), 172; Brian Linn, *Elvis's Army: Cold War GIs and the Atomic Battlefield* (Cambridge, MA: Harvard University Press, 2016), 40; Robert T. Davis, *The Challenge of Adaptation: The US Army in the Aftermath of Conflict, 1953–2000* (Fort Leavenworth, KS: CSI Press, 2008), 37n55.

4. Frank G. Hoffman, *Mars Adapting: Military Change during War* (Annapolis, MD: Naval Institute Press, 2021), 235.

5. Robert Palmer, Bell I. Wiley, and William R. Keast, *The Procurement and Training of Ground Combat Troops* (Washington, DC: GPO, 1948), 13–22, 492; and Marc R. DeVore, *When Failure Thrives: Institutions and the Evolution of Postwar Airborne Forces* (Fort Leavenworth, KS: Army Press, 2015), 32, 59; James M. Gavin, interview by Donald Andrews and Charles Ferguson, 1975, box 1: Oral History Transcripts, James M. Gavin Papers, USAHEC.

6. Hoffman, *Mars Adapting*, 235.

7. Roger Beaumont, *Military Elites: Special Fighting Units in the Modern World* (New York: Bobbs-Merrill, 1974), 103–12; Bernard B. Fall, *Hell in a Very Small Place: The Siege of Dien Bien Phu* (Philadelphia: Lippincott, 1967), 177; "Paratroopers in Algeria Irked by Political Weakness in Paris," *New York Times*, May 16, 1958. See Charles R. Schrader, *The First Helicopter War: Logistics and Mobility in Algeria, 1954–1962* (Westport, CT: Praeger, 1999); DeVore, *When Failure Thrives*, 41; Steven J. Zaloga, *Inside the Blue Berets: A Combat History of Soviet and Russian Airborne Forces, 1930–1995* (Novato, CA: Presidio, 1995), 117–63;

Jacob W. Kipp, "The Political Ballet of General Aleksandr Lebed," *Problems of Post-communism* 43, no. 4 (July 1, 1996): 43-53, https://doi.org/10.1080/10758 216.1996.11655688.

8. Allan D. English, *Understanding Military Culture: A Canadian Perspective* (Montreal: McGill-Queen's University Press, 2004), 10; Peter R. Mansoor and Williamson Murray eds., *The Culture of Military Organizations* (Cambridge: Cambridge University Press, 2019), 1; Clifford Geertz, "Religion as a Cultural System," in *The Interpretation of Cultures* (New York: Basic Books, 1973), 89; Edgar H. Schein and Peter Schein, *Organizational Culture and Leadership* (Hoboken, NJ: John Wiley, 2016), 87-102, 111.

9. Schein and Schein, *Organizational Culture and Leadership*, 2, 87-102, 111.

10. Schein and Schein, 2, 291-317, 181.

11. Schein and Schein, 152; English, *Understanding Military Culture*, 15, 23.

12. Wayne E. Lee, *Barbarians and Brothers: Anglo-American Warfare, 1500–1865* (New York: Oxford University Press, 2011), 6; John Lynn, *Battle: A History of Combat and Culture* (Boulder, CO: Westview, 2003), xx; Isabel Hull, *Absolute Destruction: Military Culture and the Practices of War in Imperial Germany* (Ithaca, NY: Cornell University Press, 2004), 97-98, 92; Wayne E. Lee, "Mind and Matter—Cultural Analysis in American Military History: A Look at the State of the Field," *Journal of American History* 93, no. 4 (2007): 1135-36, https://doi.org/10.2307/25094598; Wayne E. Lee, ed., *Warfare and Culture in World History* (New York: NYU Press, 2011), 3; Peter H. Wilson, "Defining Military Culture," *Journal of Military History* 72 (2008): 18; Tami Davis Biddle, *Rhetoric and Reality in Air Warfare: The Evolution of British and American Ideas about Strategic Bombing, 1914–1945* (Princeton, NJ: Princeton University Press, 2002), 6; Eric Michael Burke, *Soldiers from Experience: The Forging of Sherman's Fifteenth Army Corps, 1862–1863* (Baton Rouge: LSU Press, 2022).

13. Heather P. Venable, *How the Few Became the Proud: Crafting the Marine Corps Mystique, 1874–1918* (Annapolis, MD: Naval Institute Press, 2019); Aaron B. O'Connell, *Underdogs: The Making of the Modern Marine Corps* (Cambridge, MA: Harvard University Press, 2012); Craig M. Cameron, *American Samurai: Myth, Imagination, and the Conduct of Battle in the First Marine Division, 1941–1951* (New York: Cambridge University Press, 1994), 15; see also Allen Millett, "The US Marine Corps, 1973-2017: Cultural Preservation in Every Place and Clime," in Mansoor and Murray, *Culture of Military Organizations*, 379, 391.

14. Malcom Gladwell, *The Bomber Mafia: A Dream, a Temptation, and the Longest Night of the Second World War* (New York: Little, Brown, 2021); Melvin G. Dealie, *Always at War: Organizational Culture in Strategic Air Command, 1946–62* (Annapolis, MD: Naval Institute Press, 2018); Michael W. Hankins, *Flying Camelot: The F-15, the F-16, and the Weaponization of Fighter Pilot Nostalgia* (Ithaca, NY: Cornell University Press, 2021). For more on strategic bombardment see John Curatola, *Bigger Bombs for a Brighter Tomorrow: The Strategic Air Command and American War Plans at the Dawn of the Atomic Age, 1945–1950* (Jefferson, NC: McFarland, 2015); Edward Kaplan, *To Kill Nations: American Strategy in the Air-Atomic Age and the Rise of Mutually Assured Destruction* (Ithaca, NY: Cornell University Press, 2015). For more on strategic mobility see Keith Hutcheson, *Air Mobility: The Evolution of Global Reach* (Vienna, VA: Point One, 1999), and Robert

C. Owen, *Air Mobility: A Brief History of the American Experience* (Washington, DC: Potomac Books, 2013).

15. For more on World War II airborne operations see James A. Huston, *Out of the Blue: US Army Airborne Operations in World War II* (West Lafayette, IN: Purdue University Press, 1972); Clay Blair, *Ridgway's Paratroopers: The American Airborne in World War II* (Garden City, NY: Dial, 1985); Gerald Devlin, *Paratrooper! The Saga of the US Army and Marine Parachute and Glider Combat Troops during World War II* (New York: St. Martin's, 1986); William B. Breuer, *Geronimo! American Paratroopers in World War II* (New York: St. Martin's, 1992); E. M. Flanagan, *Airborne: A Combat History of American Airborne Forces* (New York: Presidio, 2002); Mitchell Yockelson, *The Paratrooper Generals: Matthew Ridgway, Maxwell Taylor, and the American Airborne from D-Day through Normandy* (Guilford, CT: Stackpole Books, 2020). For more on the 1950s army see Bacevich, *Pentomic Era*; Ingo Trauschweizer, *The Cold War US Army: Building Deterrence for Limited War* (Lawrence: University Press of Kansas, 2008); Linn, *Elvis's Army*; Donald Carter, *The US Army before Vietnam* (Washington, DC: CMH, 2015); Kalev I. Sepp, "The Pentomic Puzzle: The Influence of Personalities and Nuclear Weapons on US Army Organization 1952–1958," *Army History* 51 (Winter 2001): 9; John J. Midgley Jr., *Deadly Illusions: Army Policy for the Nuclear Battlefield* (Boulder, CO: Westview, 1986).

1. The Birth of American Airborne Culture

1. Donald L. Deam, *General Toothpick: The WWII Memoirs of 1st Sgt. Donald L. Deam* (Fort Campbell, KY: 101st Airborne Division, 2008), 6.

2. "U.S. Trains Parachutists," *Life*, August 19, 1940; "U.S. Trains More Parachute Troops," *Life*, May 12, 1941; Leslie Goodwins, dir., *Parachute Battalion* (RKO Pictures, 1941); Edward M. Coffman, *The Regulars: The American Army, 1898–1941* (Cambridge, MA: Harvard University Press, 2004), 403. The 501st Parachute Battalion, including William P. Yarborough, provided extras for the film and conducted all live jumps.

3. Walter Duranty, "Soviets Initiate Parachute Attack," *New York Times*, September 16, 1935; Zaloga, *Inside the Blue Berets*, 13–14; James S. Herndon and Joseph O. Baylen, "Col. Philip R. Faymonville and the Red Army, 1934–43," *Slavic Review* 34, no. 3 (September 1975): 488, https://doi.org/10.2307/2495561; Franz Kurowski, *Jump into Hell: German Paratroopers in World War II* (Mechanicsburg, PA: Stackpole Books, 2010), 165.

4. DeVore, *When Failure Thrives*, 22; William P. Yarborough oral history interview, 1975, box 1, William P. Yarborough Papers, USAHEC; James M. Gavin, *On to Berlin: A Fighting General's True Story of Airborne Combat in World War II* (New York: Bantam Books, 1981), 2.

5. The War Department ordered the formation of the first three parachute regiments on January 30, 1942, and the Provisional Parachute Group became the Airborne Command on March 21, 1942; see Memorandum for General Bull, Assistant Chief of Staff, G-3, Subj: Air-borne Divisions, April 4, 1942, in *The Papers of Dwight D. Eisenhower: The War Years*, vol. 1, ed. Alfred D. Chandler (Baltimore: Johns Hopkins University Press, 1970), 226; DeVore, *When Failure*

Thrives, 57; Gordon Rottman, *US Army Airborne, 1940–1990* (New York: Osprey, 1990), 7.

6. A. D. Rathbone IV, *He's in the Paratroops Now* (New York: Robert M. Mc-Bride, 1943), 24; Tania Chaco, "Why Did They Fight? American Airborne Units in World War II," *Defence Studies* 1, no. 3 (Autumn 2001): 71–72, https://doi.org/10.1080/714000045; Devlin, *Paratrooper!*, 49–51, 109; Cole C. Kingseed, *Conversations with Major Dick Winters: Life Lessons from the Commander of the Band of Brothers* (New York: Berkley Caliber, 2014), 49.

7. James M. Gavin oral history, 1975, section 1, pp. 11–12, box 1, James M. Gavin Papers, USAHEC (hereafter Gavin Oral History); "Paratroops "Best Outfit in the Army," Says Corporal Jim Perham, Home on Leave," *Daily Tribune* (Wisconsin Rapids, WI), September 14, 1942; Vincent J. Speranza, *NUTS! A 101st Airborne Division Machine Gunner at Bastogne* (Atlanta: Deeds, 2014), 3; Jeremy C. Holm, *When Angels Fall: From Toccoa to Tokyo* (Salt Lake City, UT: Holm, 2019, 19; Samuel A. Stouffer et al., *The American Soldier*, vol. 2, *Combat and Its Aftermath* (Princeton, NJ: Princeton University Press, 1949), 329.

8. Palmer, Wiley, and Keast, *Procurement and Training*, 14, 20, 67; Peter R. Mansoor, *The GI Offensive in Europe: The Triumph of American Infantry Divisions, 1941–1945* (Lawrence: University Press of Kansas, 1999), 41.

9. Kurt Gabel, *The Making of a Paratrooper: Airborne Training and Combat in World War II* (Lawrence: University Press of Kansas, 1990), 31–32; "Procurement of Infantry Personnel for the Parachute Battalion," memorandum, September 18, 1940, Early Airborne, box 5, WMM82.

10. US War Department, FM 31-30, *Tactics and Technique of Air-borne Troops* (Washington, DC: GPO, 1942), 31 (hereafter FM 31-30, *Tactics and Technique*); Holm, *When Angels Fall:* 26–33; Stephen E. Ambrose, *Band of Brothers: E Company, 506th Regiment, 101st Airborne from Normandy to Hitler's Eagle's Nest* (New York: Simon & Schuster, 2001), 16; William Guarnere and Edward Heffron with Robyn Post, *Brothers in Battle, Best of Friends: Two WWII Paratroopers from the Original Band of Brothers Tell Their Story* (New York: Berkley Caliber, 2007), 25; Kingseed, *Conversations with Major Dick Winters*, 170; Gideon Aran, "Parachuting," *American Journal of Sociology* 80, no. 1 (July 1974): 150, https://doi.org/10.1086/225764.

11. Gabel, *Making of a Paratrooper*, 37, 44; James B. Carlaw, oral history interview by Shaun Illingworth, July 21, 2006, Rutgers Oral History Archives, New Brunswick, NJ, John Foster Magill Oral History Transcript (AFC/2001/001/80043), LOC.

12. Mark J. Alexander and John Sparry, *Jump Commander: In Combat with the 505th and 508th Parachute Infantry Regiments, 82nd Airborne Division in World War II* (Philadelphia: Casemate, 2018), 44; Donald R. Burgett, *Currahee! A Screaming Eagle at Normandy* (Novato, CA: Presidio, 1967), 8; Matthew B. Ridgway as told to Harold H. Martin, *Soldier: The Memoirs of Matthew B. Ridgway* (New York: Harper, 1956), 53; Richard Holmes, *Acts of War: The Behavior of Men in Battle* (New York: Free Press, 1986), 48.

13. Ridgway and Martin, *Soldier*, 51–54.

14. Parachute Training Course Curriculum, Early Airborne, box 6, WMM82.

15. Patent number 134963, US Patent Office, June 19, 1942, Early Airborne, box 17, WMM82; Guarnere, Heffron, and Post, *Brothers in Battle, Best of Friends,* 99; Peter Harclerode, *Wings of War: Airborne Warfare 1918–1945* (London: Cassell, 2005), 33–34.

16. "Approve New Emblem for Parachutists," *Schenectady Gazette,* July 2, 1941; Stouffer et al., *American Soldier,* 2:329; Coffman, *Regulars,* 403; Guy LoFaro, *The Sword of St. Michael: The 82nd Airborne Division in World War II* (Cambridge, MA: Da Capo, 2011), 19; Gerald Astor, *Battling Buzzards: The Odyssey of the 517th Parachute Regimental Combat Team 1942–1945* (New York: Dell, 1993), 9.

17. Francis L. Sampson, *Look Out Below! A Story of the Airborne by a Paratrooper Padre* (Sweetwater, TN: 101st Airborne Division Association, 1989), 16; McNair quoted in Breuer, *Geronimo!,* 9. For more on fighting see Spencer F. Wurst and Gayle Wurst, *Descending from the Clouds: A Memoir of Combat in the 505 Parachute Infantry Regiment, 82nd Airborne Division* (Havertown, PA: Casemate, 2004), 51; John Foster Magill Oral History; Gavin Oral History, section 1, p. 15, USAHEC; for more on the downside of airborne culture see R. F. M. Williams, "'Our Problem Children': Masculinity and Its Discontents in American Parachute Units in World War II," *Journal of Military History* 87, no. 3 (July 2023): 675–702.

18. Bob Bearden, *To D-Day and Back: Adventure with the 507th Parachute Infantry Regiment and Life as a World War II POW* (Minneapolis, MN: Zenith, 2007), 23; "Prop Blast Ceremony Rules," 1951, WWII, box 17, WMM82; Devlin, *Paratrooper!,* 90–91; T. Michael Booth and Duncan Spencer, *Paratrooper: The Life of Gen. James M. Gavin* (New York: Simon & Schuster, 1994), 77. For more on songs see Wurst and Wurst, *Descending from the Clouds,* 44; Devlin, *Paratrooper!,* 124.

19. Alexander and Sparry, *Jump Commander,* 49; Ambrose, *Band of Brothers,* 34; Kingseed, *Conversations with Major Dick Winters,* 133.

20. Ridgway, "Outline," n.d., box 58, Official Papers, 82nd Airborne Division, XVIII Airborne Corps, 1942–1945, Ridgway Papers, USAHEC; Beaumont, *Military Elites,* 6; Holmes, *Acts of War,* 48.

21. FM 31-30, *Tactics and Technique,* 32, 49; John T. Ellis, *The Airborne Command and Center,* Army Ground Forces Historical Section Study No. 25 (Washington, DC: Army Ground Forces, 1946), 9.

22. AGF M/S, TRC to CoS, June 22, 1942, "Policy re Training of Airborne Troops, appendix 23, in Ellis, *Airborne Command and Center,* 132; Kent R. Greenfield, Robert R. Palmer, and Bell I. Wiley, *The Organization of Ground Combat Troops* (Washington, DC: Department of the Army, 1947), 340; "Organization, Equipment, and Tactical Employment of the Airborne Division," p. 9, Study No. 16, Reports of the General Board, US Forces, European Theater, CARL; Airborne Division Table of Organization (T/O 71) 15 Oct 42, WWII, box 43, WMM82; Ridgway to Marshall, November 1, 1944, George C. Marshall Papers, Pentagon Office Collection, Selected Materials, GCMRL; Marshall to Ridgway, December 18, 1944, Marshall Papers, Pentagon Office Collection, GCMRL; Ridgway and Martin, *Soldier,* 126.

23. Jack Thompson, "Lack of Planes Delays Training of Paratroops," *Chicago Daily Tribune,* July 14, 1941; "Allocation of Transport Planes for U.S.S.R.,"

Meeting Minutes, 16th Meeting of the Combined Chiefs of Staff, April 21, 1942, cited from *Minutes of Meetings of the Combined Chiefs of Staff, Post-Arcadia,* vol. 1 (Washington, DC: Joint History Office, 2003), 130; Christopher R. Gabel, *The U.S. Army GHQ Maneuvers of 1941* (Washington, DC: GPO, 1992), 41.

24. Commander, 82nd A/B Division, to Commander-in-Chief, Allied Force, memorandum, "Summary of Principles Covering Use of the Airborne Division," November 27, 1943, Personal Papers, box 46, Matthew B. Ridgway, WMM82; Blair, *Ridgway's Paratroopers,* 31, 49-51; Ellis, *Airborne Command and Center,* 10.

25. FM 31-30, *Tactics and Technique,* 61; Gavin, *On to Berlin,* 2.

26. Briton Cooper Busch, *Bunker Hill to Bastogne: Elite Forces and American Society* (Dulles, VA: Potomac Books, 2006), 167; HQ Airborne Command to CGs 82nd and 101st Airborne Divisions, memorandum, "Training Directive," November 4, 1942, appendix 25, in Ellis, *Airborne Command and Center,* 137-42; Gabel, *Making of a Paratrooper,* 71; Wurst and Wurst, *Descending from the Clouds,* 46.

27. HQ Airborne Command to CGs 82nd and 101st Airborne Divisions, memorandum, "Training Directive," November 4, 1942, appendix 25, in Ellis, *Airborne Command and Center,* 137-42; Burgett, *Currahee,* I; Hoffman, *Mars Adapting,* 235.

28. Larry Alexander, *Biggest Brother: The Life of Major Dick Winters, the Man Who Led the Band of Brothers* (New York: Penguin, 2005), 41; Gavin Oral History, section 1, 7-9, USAHEC; Henry Langrehr and Jim DeFelice, *Whatever It Took: An American Paratrooper's Extraordinary Memoir of Escape, Survival, and Heroism in the Last Days of World War II* (New York: William Morrow, 2020), 40.

29. William Ryder to James Gavin, May 17, 1941, box 7, Gavin Papers; Gavin, *On to Berlin,* 2; James M. Gavin, *War and Peace in the Space Age* (New York: Harper, 1958), 46.

30. Maxwell D. Taylor, *Swords and Plowshares* (New York: W. W. Norton, 1972), 43-45; John M. Taylor, *General Maxwell Taylor: The Sword and the Pen* (New York: Doubleday, 1989), 43.

31. FM 31-30, *Tactics and Technique,* 32.

32. Irvin Williams Seelye Oral History Transcript (AFC/2001/001/02027), LOC; Taylor, *Swords and Plowshares,* 47; Jörg Muth, *Command Culture: Officer Education in the U.S. Army and the German Armed Forces, 1901–1940* (Denton: University of North Texas Press, 2011), 102; Stouffer et al., *American Soldier,* 2:118-25.

33. Marshall Andrews, "Our New Army," *Washington Post,* May 9, 1942; John Thompson, "Chute Troopers' Morale Goes Up with Each Fall," *Chicago Daily Tribune,* February 2, 1942; Lamar Q. Ball, "Today's Paratrooper Scorns Fatalism Fad," *Atlanta Constitution,* November 27, 1942.

34. Taylor, *Swords and Plowshares,* 45; Andrew Carrico, as quoted in Holm, *When Angels Fall,* 50; see also p. 30.

35. Gavin Oral History, section 1, p. 15, USAHEC; Theodore Wilson in Gabel, *Making of a Paratrooper,* xi-xii; David Bergman, Marie Gustafsson Senden, and Erik Berntson, "Preparing to Lead in Combat: Development of Leadership Self-Efficacy by Static-Line Parachuting," *Military Psychology* 31, no. 6 (October 2019): 489, https://doi.org/10.1080/08995605.2019.1670583; on

division patches see Peter S. Kindsvatter, *American Soldiers: Ground Combat in the World Wars, Korea, and Vietnam* (Lawrence: University Press of Kansas, 2003), 133, 134; see also Mansoor, *GI Offensive in Europe*, 264.

36. HQ Airborne Command to CGs 82nd and 101st Airborne Divisions, "Training Directive," November 4, 1942, appendix 25, in Ellis, *Airborne Command and Center*, 137–42.

37. Ridgway and Martin, *Soldier*, 7; FM 31–30, *Tactics and Technique*, 30, 32.

38. Alexander and Sparry, *Jump Commander*, 50; Gavin, *On to Berlin*, 2–3.

39. Gavin Oral History, section 1, p. 14, USAHEC; Gavin, *On to Berlin*, 3.

40. John A. Lynn, *Bayonets of the Republic: Motivation and Tactics in the Army of Revolutionary France, 1791–94* (Boulder, CO: Westview, 1996), 262–63; Holger Herwig, "The Dynamics of Necessity: German Military Policy during the First World War," in *Military Effectiveness*, vol. 1, *The First World War*, ed. Allan R. Millett and Williamson Murray (Cambridge: Cambridge University Press, 2010), 101. For a robust exploration of German tactical development see Bruce I. Gudmundsson, *Stormtroop Tactics: Innovation in the German Army, 1914–1918* (Westport, CT: Praeger, 1995).

41. For more on infantry evolution since at least the mid-eighteenth century see Burke, *Soldiers from Experience*, 269–72; John Grenier, *The First Way of War: American War Making on the Frontier, 1607–1814* (New York: Cambridge University Press, 2010); and John A. English and Bryce I. Gudmundsson, *On Infantry* (Westport, CT: Praeger, 1994). For a broader perspective see Clifford J. Rogers, ed., *The Military Revolution Debate: Readings on the Military Revolution of Early Modern Europe* (New York: Routledge, 1995).

42. Mansoor, *GI Offensive*, 31; Keast, Palmer, and Wiley, *Procurement and Training*, 596–97; Guarnere, Heffron, and Post, *Brothers in Battle*, 43.

43. FM 31–30, *Tactics and Technique*, 32; Richard Seitz, as quoted in Astor, *Battling Buzzards*, 28–29.

44. Ellis, *Airborne Command and Center*, 9; Langrehr and DeFelice, *Whatever It Took*, 38.

45. Palmer, Wiley, and Keats, *Procurement and Training*, 442–69; Mansoor, *GI Offensive*, 82–83; W. Forrest Dawson, ed., *Saga of the All American* (Atlanta: Albert Love, 1946).

46. Gabel, *Making of a Paratrooper*, 134; Guarnere, Heffron, and Post, *Brothers in Battle*, xx; Dick Eaton, as quoted in Astor, *Battling Buzzards*, 395.

47. Robert J. MacCoun, Elizabeth Kier, and Aaron Belkin, "Does Social Cohesion Determine Motivation in Combat? An Old Question with an Old Answer," *Armed Forces and Society* 32, no. 1 (2005): 1–9, https://www.jstor.org/stable/48608737; Anthony King, *The Combat Soldier: Infantry Tactics and Cohesion in the Twentieth and Twenty-First Centuries* (New York: Oxford University Press, 2013), 25–34, 350; William Cockerham and Lawrence E. Cohen, "Volunteering for Foreign Combat Missions: An Attitudinal Study of U.S. Army Paratroopers," *Pacific Sociological Review* 24, no. 3 (July 1981): 351, https://doi.org/10.2307/1388810; Sheldon G. Levin, *Mathematical Models for Prediction of Neuropsychiatric and Other Non-battle Casualties in High Intensity Combat* (Aberdeen Proving Ground, MD: US Army Ballistic Research Laboratory, 1986), 15.

48. Blair, *Ridgway's Paratroopers*, 27.

49. John Foster Magill Oral History; Sampson, *Look Out Below!*, 15; Jack P. Nix, "505 Parachute Infantry Regiment (a Legacy of Lessons)," US Army War College Military Studies Program Paper, Carlisle Barracks, PA, 1989, 7.

50. Rathbone, *He's in the Paratroops Now*, 109; Huston, *Out of the Blue*, 49.

51. Palmer, Wiley, and Keast, *Procurement and Training*, 258; Beaumont, *Military Elites*, 97, 102. Although most paratroopers attended the Parachute School, some made their first jumps in combat. See Werner T. Angress, *Witness to the Storm: A Jewish Journey from Nazi Berlin to the 82nd Airborne, 1920–1945* (Bloomington: Indiana University Press, 2012), 257–62.

52. William Cockerham, "Selective Socialization: Airborne Training as a Status Passage," *Journal of Political and Military Sociology* 1 (Fall 1973): 215–29, http://www.jstor.org/stable/45293603.

2. World War II and the Foundation of the Airborne Mafia

1. Maj. Gen. William M. Miley to Lt. Col. J. B. Shinberger, October 13, 1943, box 2, folder 20, William M. Miley Papers, USAHEC (hereafter Miley Papers).

2. WSEG, "A Historical Study of Some World War II Airborne Operations," WSEG Staff Study No. 3, 1951, 139; Rick Atkinson, *An Army at Dawn* (New York: Henry Holt, 2002), 90; Edson Raff, *We Jumped to Fight* (New York: Eagle Books, 1944), 64; William Yarborough, *Bail Out over North Africa: America's First Combat Parachute Missions, 1942* (Williamstown, NJ: Phillips, 1979), 92; Blair, *Ridgway's Paratroopers*, 66–67.

3. Yarborough, *Bail Out over North Africa*, 97, 101, 126; Raff, *We Jumped to Fight*, 78–103, 138; Huston, *Out of the Blue*, 152–53.

4. 505th Parachute Infantry AAR, memorandum to CG, 82nd Airborne Division, August 14, 1943, World War II Operational Documents, CARL; Allied Force Hdqtrs, N Africa—Rep of Allied Force Airborne Board on Op "HUSKY" Department of the Navy, Office of the Chief of Naval Operations, Intelligence Division, World War II War Diaries, RG 38: Records of the Office of the Chief of Naval Operations, 1875–2006, NARA II; H.Q., 82nd A/B Division, "The 82d Airborne Division in Sicily and Italy," November 1, 1945, 5–6, World War II Operational Documents, CARL.

5. Ridgway and Martin, *Soldier*, 68–69; Gavin, *On to Berlin*, 19.

6. Lt. Col. Charles Billingslea, "Report of A/B Operations, 'Husky' and 'Bigot,'" August 15, 1943, A/B Overseas Rpts, p. 11, World War II Operational Documents, CARL; Albert Garland, Howard Smyth, and Martin Blumenson, *Sicily and the Surrender of Italy: The Mediterranean Theater of Operations* (Washington, DC: GPO, 1965), 182; Taylor, *Swords and Plowshares*, 50; Alexander and Sparry, *Jump Commander*, 87.

7. Alexander and Sparry, *Jump Commander*, 81; William B. Breuer, *Drop Zone Sicily: Allied Airborne Strike, July 1943* (Novato, CA: Presidio, 1983), 20–21.

8. Billingslea, report, August 15, 1943, p. 10, CARL; Gavin, *On to Berlin*, xiii; 82nd Airborne Division in Sicily and Italy, HQ, 82nd A/B Division, November 1, 1945, p. 13, World War II Operational Documents, CARL; Ridgway and Martin, *Soldier*, 69–70; see James M. Gavin, *Airborne Warfare* (Washington, DC: Infantry Journal Press, 1947), 16.

9. Gavin, *On to Berlin*, 90. For Patton comments see report of Maj. Gen. F. A. M. Browning, July 24, 1943, World War II Operational Documents, CARL.

10. Dwight D. Eisenhower, *Crusade in Europe* (New York: Doubleday, 1948), 173; Eisenhower to Marshall, September 20, 1943, in *Papers of Dwight D. Eisenhower: The War Years*, 3:1440; Swing in Huston, *Out of the Blue*, 164.

11. Gavin, *On to Berlin*, 52; Edwin M. Sayre, "The Operations of Company A, 505th Parachute Infantry (82nd Airborne Division), Airborne Landings in Sicily, 9–24 July 1943 (Sicily Campaign) Personal Experiences of a Company Commander" (Fort Benning, GA, 1947), 21.

12. War Department, Training Circular No. 113, *Employment of Airborne and Troop Carrier Forces*, October 9, 1943, 2, 4; Eugene G. Piasecki, "The Knollwood Manuever: The Ultimate Test," *Veritas* 4, no. 1 (2008); E. M. Flanagan, *Airborne*, 103.

13. "Suitability of the Planned Operations for Execution by Airborne Troops," memorandum, HQ, 82nd Airborne Division, October 25, 1943, and letter from Lt. Gen. Mark Clark to Maj. Gen. Matthew Ridgway, September 13, 1943, World War II, box 6, HQ Paperwork, WMM82; Yarborough Oral History, 47–48, USAHEC; Martin Blumenson, *Salerno to Cassino* (Washington, DC: GPO, 1969), 131.

14. Mark Clark, *Calculated Risk* (New York: Harper Collins, 1950), 169; Ridgway and Martin, *Soldier*, 86; Chester G. Starr, *From Salerno to the Alps: A History of the Fifth Army, 1943–1945* (Washington, DC: Infantry Journal Press, 1948), 29.

15. "Organization, Equipment, and Tactical Employment of the Airborne Division," 4, Study No. 16, Reports of the General Board, U.S. Forces, ETO, CARL; Gavin, *On to Berlin*, 133; Lynn "Buck" Compton with Marcus Brotherton, *Call of Duty: My Life before, during, and after the Band of Brothers* (New York: Berkley Caliber, 2008), 14; Dwayne Burns with Leland Burns, *Jump into the Valley of the Shadow: The World War II Memories of a Paratrooper in the 508th P.I.R., 82nd Airborne Division* (Philadelphia: Casemate, 2006), 40.

16. Field Order No. 6, 82nd Airborne Division, May 28, 1944, World War II Operational Documents, CARL; Maj. Salve H. Matheson, "The Operations of the 506th Parachute Infantry (101st Airborne Division) in the Normandy Invasion, 5–8 June 1944 (Normandy Campaign), Personal Experience of a Regimental Staff Officer," Fort Benning, GA, 1949, p. 7, DRL; "Organization, Equipment, and Tactical Employment of the Airborne Division," 3; Taylor, *Swords and Plowshares*, 73.

17. Gavin, *Airborne Warfare*, 63; Eugene Andrew Drance Collection (AFC/2001/001/85477), Veterans History Project, American Folklife Center, LOC; for Taylor quote see Taylor, *Swords and Plowshares*, 80–82; Huston, *Out of the Blue*, 182.

18. 82nd Airborne Division in Normandy, France—Operation Neptune, U.S. Army Unit Records, box 6, DDEL; "Organization, Equipment, and Tactical Employment of the Airborne Division," 4; Ridgway to Eisenhower, memorandum, July 10, 1944, folder Ridgway, Matthew B., box 98, Pre-presidential Papers, DDEL; 505th PIR Regimental S3 journal, Normandy, WWII, box 30, WMM82; After Action Report, "Operations of the 507th RCT following the Drop," WWII, box 33, WMM82; Huston, *Out of the Blue*, 183.

19. Maj. Knut H. Raudstein, "The Operation of the 506th Parachute Infantry (101st Airborne Division) in the vicinity of Carentan, June 6–8, 1944 (Normandy Campaign)," Infantry School student paper, Fort Benning, GA, 1948, 41; Gavin Oral History, section 2, p. 8, USAHEC; Burns, *Jump into the Valley of the Shadow*, 59; "Organization, Equipment, and Tactical Employment of the Airborne Division," 4. For more on La Fière see Robert M. Murphy, *No Better Place to Die: The Battle for La Fière Bridge* (Havertown, PA: Casemate, 2009); Letter to Combined Chiefs of Staff, Cable #90024, Top Secret, June 8, 1944, *Papers of Dwight D. Eisenhower: The War Years*, 3:1736–39.

20. Omar N. Bradley, *A Soldier's Story* (New York: Henry Holt, 1951), 235; Alexander and Sparry, *Jump Commander*, 233; Blair, *Ridgway's Paratroopers*, 295–96.

21. WSEG Staff Study No. 3, p. 21; Gavin, *Airborne Warfare*, 66; Matheson, "Operations of the 506th Parachute Infantry in the Normandy Invasion," 7.

22. Record of Debriefing Conference, Operation Neptune, August 13, 1944, 13–14, Personal Papers, box 13, WMM82; Gavin, *On to Berlin*, 133.

23. Allan R. Millett, "The United States Armed Forces in the Second World War," in *Military Effectiveness*, vol. 3, *The Second World War*, ed. Allan R. Millett and Williamson Murray (New York: Cambridge University Press, 2010), 68; Huston, *Out of the Blue*, 191–92.

24. Blumenson, *Salerno to Cassino*, 77; Gavin, *Airborne Warfare*, 28.

25. Clark, *Calculated Risk*, 161; Gavin, *Airborne Warfare*, 28; Lt. Gen. Mark Clark to Maj. Gen. Matthew Ridgway, September 13, 1943, World War II, box 6, WMM82.

26. Mansoor, *GI Offensive in Europe*, 114; Yarborough Oral History, 40; Ridgway and Martin, *Soldier*, 85.

27. 504th Staff Journal, Italy, September 1943, box 261, Veteran Survey Collection, USAHEC; Gavin, *Airborne Warfare*, 29; Clark, *Calculated Risk*, 167; HQ, 82nd Airborne Division, "The 82d Airborne Division in Sicily and Italy," 49, box 12346, RG 407, NARA; Blumenson, *Salerno to Cassino*, 127; Gavin, *On to Berlin*, 72, 74.

28. HQ, 82nd Airborne Division, "82d Airborne Division in Sicily and Italy," 50; Irvin Seelye Oral History (AFC/2001/001/02027), Veterans History Project, American Folklife Center, LOC; Mansoor, *GI Offensive in Europe*, 114.

29. Gavin, *Airborne Warfare*, 31.

30. William A. Wellman, dir., *Battleground* (MGM, 1949); Steven Spielberg and Tom Hanks, prods., *Band of Brothers*, episode 6, "Bastogne," dir. David Leland, aired October 7, 2001, HBO; Maxwell D. Taylor interview with Richard A. Manion, section 1, pp. 3–4, November 10, 1972, Senior Officers Debriefing Program, Maxwell Taylor Papers, USAHEC (hereafter Taylor Oral History); John D. McKenzie, *On Time, on Target: The World War II Memoir of a Paratrooper in the 82d Airborne* (Novato, CA: Presidio, 2000), 51.

31. Ridgway to Gavin, Ridgway to Taylor, November 14, 1944, box 21, Ridgway Papers; James Megellas, *All the Way To Berlin: A Paratrooper at War in Europe* (New York: Presidio, 2003), 180; Taylor to Ridgway, November 7, 1944, box 21, Ridgway Papers; Guarnere, Heffron, and Post, *Brothers in Battle*, 153; Speranza, *NUTS!*, 42.

32. Taylor, *Swords and Plowshares*, 97; Maxwell Taylor interview with Forrest C. Pogue, July 16, 1959, 9, Pogue interviews, GCMRL; Gavin, *On to Berlin*, 228.

33. Ridgway and Martin, *Soldier*, 114; Bradley, *Soldier's Story*, 256, 467, 480; LoFaro, *Sword of St. Michael*, 445; Ozzie Schock to Al Ireland, August 11, 1997, box 17, Gavin Papers, WMM82; McKenzie, *On Time, on Target*, 67.

34. Phil Nordyke, *All American, All the Way: The Combat History of the 82nd Airborne Division in World War II* (St. Paul, MN: Zenith, 2005), 627; XVIII Airborne Corps (@18airbornecorps), Twitter Thread, December 23, 2020, 6:49 a.m., https://twitter.com/18airbornecorps/status/1341727627359498242; US Army CMH, "23 December 1944–Wednesday Wisdom–Battle of the Bulge," Facebook, December 23, 2020, https://www.facebook.com/armyhistory/photos/a.410473127852/10159179358687853/?type=3.

35. Ridgway and Martin, *Soldier*, 113–16; Stephen R. Taaffe, *Marshall and His Generals: U.S. Army Commanders in World War II* (Lawrence: University Press of Kansas, 2011), 271–72.

36. Ralph Mitchell, *The 101st Airborne Division's Defense of Bastogne* (Fort Leavenworth, KS: CSI Press, 1986), 9–10; Mark Bando, *101st Airborne: The Screaming Eagles in World War II* (St. Paul, MN: Zenith, 2007), 187–88; Taylor, *Swords and Plowshares*, 98; Speranza, *NUTS!*, 46.

37. Harry W. O. Kinnard Oral History, 1983, tape k50, pp. 44–46, box 1, Harry W. O. Kinnard Papers, USAHEC (hereafter Kinnard Oral History); Bradley, *Soldier's Story*, 472.

38. "Tank Leader Tells of Bastogne Fight," *New York Times*, June 25, 1945; George S. Patton, *War as I Knew It* (Boston: Houghton Mifflin, 1947), 205; John S. D. Eisenhower, *The Bitter Woods: The Battle of the Bulge* (New York: Da Capo, 1969), 345; "U.S. Tanks Smash into Bastogne, Break Siege," *New York Times*, December 28, 1944; Letter to Omar Bradley and Jacob Devers, March 12, 1945, *Papers of Dwight D. Eisenhower: The War Years*, 4:2523–24; "Wounded Pleaded for Bastogne Role," *New York Times*, January 3, 1945; Cable to Combined Chiefs of Staff, Cable S 75872, Top Secret, January 20, 1945, *Papers of Dwight D. Eisenhower: The War Years*, 4:2447.

39. Taylor's other great disappointment was that the planned airborne assault on Rome was not feasible; see Taylor Oral History, section 6, p. 11; Kinnard Oral History, tape K50, p. 52; Eugene Andrew Drance Collection (AFC/2001/001/85477), Veterans History Project, American Folklife Center, LOC.

40. James Huston, "Thoughts on the American Airborne Effort in World War II," *Military Review*, May 1951, 20.

41. For more on adaptation in combat see Hoffman, *Mars Adapting*, and David Barno and Nora Benashel, *Adaptation under Fire: How Militaries Change in Wartime* (New York: Oxford University Press, 2020); Williamson Murray, *Adaptation in War: With Fear of Change* (New York: Cambridge University Press, 2011); Huston, *Out of the Blue*, 187; Millett, "United States Armed Forces in the Second World War," 68.

42. M. B. Ridgway to CG, FIFTH ARMY, memorandum, Subj: "Operations," August 20, 1943, box 19, Ridgway Papers; Army Ground Forces Historical

Section, Study No. 25, 1946, 6; "Organization, Equipment, and Tactical Employment of the Airborne Division," 6–7, 10.

43. Ridgway and Martin, *Soldier*, 12, 62; Sayre, "Operations of Company A," 22; "Organization, Equipment, and Tactical Employment of the Airborne Division," 9.

44. Mansoor, *GI Offensive in Europe*, 257; Taylor, *Swords and Plowshares*, 50; Maj. Gen. J. M. Swing, memorandum, "Analysis of the Airborne Participation in Operation 'Husky,'" Am HQ Force 141, July 16, 1943, World War II Operational Documents, CARL; Gavin, *Airborne Warfare*, 163.

45. Airborne Division Table of Organization (T/O 71) 15 Oct 42 and 24 Feb 44, WWII, box 43, WMM82; Huston, *Out of the Blue*, 187; Huston, "Thoughts on the American Airborne Effort," 23.

46. Ridgway to Marshall, November 1, 1944, and Marshall to Ridgway, December 18, 1944, George C. Marshall Papers, Pentagon Office Collection, Selected Materials, GCMRL; Ridgway and Martin, *Soldier*, 126; Col. Edwin W. Chamberlain to Maj. Gen. Maxwell D. Taylor, memorandum, December 11, 1944, box 21, Ridgway Papers; Millet, "United States Armed Forces in the Second World War," 71.

47. Taylor Oral History, section 2, pp. 29–30; Huston, "Thoughts on the American Airborne Effort," 23.

48. James Fenelon, *Four Hours of Fury: The Untold Story of World War II's Largest Airborne Operation and the Final Push into Nazi Germany* (New York: Scribner, 2019), 135. Glider pilots in the British airborne were formally assigned to their own infantry companies once on the ground.

49. Phil Nordyke, *All American*, 105; Gavin, *Airborne Warfare*, 61–62; HQ, 505th Parachute Infantry, "Air Re-Supply Plan," May 28, 1944, box 1, William E. Ekman Papers, USAHEC; Inspector General to CG, 82nd Airborne Division, "Glider Operation NEPTUNE," August 4, 1944, box 21, Ridgway Papers; Teddy H. Sanford remarks in Record of Debriefing Conference, Operation Neptune, August 13, 1944, Personal Papers, box 13, WMM82; Eisenhower to Marshall, cable, and Ridgway to Marshall, letter, July 23, 1944, box 20, Ridgway Papers.

50. Taylor, *Swords and Plowshares*, 82; Wurst and Wurst, *Descending from the Clouds*, 128–29; James E. Mrazek, *Airborne Combat: The Glider War / Fighting Gliders of World War II* (Mechanicsburg, PA: Stackpole Books, 2011), 148–50; 82nd Airborne Division, Action in Normandy, Report of Maj. Gen. Ridgway, WWII, box 80, Normandy Historical Data, WMM82; Fenelon, *Four Hours of Fury*, 132.

51. HQ, XVIII Airborne Corps, Report of Airborne Phase (September 17–27, 1944) Operation "Market," October 5, 1944, CARL; John L. Lowden, *Silent Wings at War: Glider Combat in World War II* (Washington, DC: Smithsonian Institution, 1992), 126–29; Mrazek, *Airborne Combat*, 212–17.

52. Gavin, *Airborne Warfare*, 137; Gerald M. Devlin, *Silent Wings: The Saga of the U.S. Army and Marine Combat Glider Pilots during World War II* (New York: St. Martin's, 1985), 301–4; HQ, 17th Airborne Division, Operation Varsity, July 1945, CARL; HQ, XVIII Corps (Airborne), Report on Operation Varsity, April 25, 1945, box 2B, Miley Papers; Mrazek, *Airborne Combat*, 248.

53. Alexander, *Jump Commander*, 184–85; Lowden, *Silent Wings at War*, 111–14.

54. War Department, Study No. 17, Types of Divisions—Post War Army, Reports of the General Board U.S. Forces, European Theater of Operations, 1945, pp. 15–16, quote on 15, CARL; *Glider Breeze*, March 14, 1946, WWII, box 13, 325th GIR, WMM82; CS 2045/8, April 26, 1951, Records of the Air Force Representative on the Joint Airborne Troop Board, RG 340, NARA; Joint Airborne Troop Board Administrative and Project Files, 1950–1954, RG 337, NARA.

55. Maxwell D. Taylor, interview with Charles R. Smith, p. 8, June 12, 1974, Maxwell Taylor Papers, USAHEC; Message to General of the Army Douglas MacArthur, Radio, August 2, 1945, Marshall Papers, Pentagon Office Collection, Selected Materials, GCMRL; War Department, Study No. 17: Types of Divisions—Post War Army, Reports of the General Board U.S. Forces, European Theater of Operations, 1945, p. 15, CARL.

56. Gavin, *On to Berlin*, 327; Flanagan, *Airborne*, 300–302; Werner T. Angress, *Witness to the Storm: A Jewish Journey from Nazi Berlin to the 82nd Airborne, 1920–1945* (Bloomington: Indiana University Press, 2019), 307; Burns, *Jump into the Valley of the Shadow*, 204; Gavin, *Airborne Warfare*, 137.

57. Correspondence regarding Inactivation of Airborne Divisions, box 37, Gavin Papers; Ridgway to Floyd Parks, September 26, 1945, box 61, Ridgway Papers; Ridgway to Chief of Staff, memorandum, "Inactivation of Airborne Divisions," October 18, 1945, and Eisenhower to Ridgway, message, November 5, 1945, folder Ridgway, box 98, Pre-presidential Papers, DDEL; Letters to Barbara, March 21, 1945, October 11, 1945, and November 12, 1945, in Barbara Gavin Fauntleroy, *The General and His Daughter: The Wartime Letters of General James M. Gavin to His Daughter Barbara* (New York: Fordham University Press, 2007); Meyer Berger, "City Millions Hail the 82d in GI Tribute," *New York Times*, January 13, 1946; Bradley Biggs, *Gavin* (Hamden, CT: Archon Books, 1980), 65.

3. The Airborne Way of War and Its Strategic Implications

1. Epigraph from Robert Haldane oral history, 1985, p. 33, box 1, Robert Haldane Papers, USAHEC (hereafter Haldane Oral History); Hamilton Howze oral history, 1973, section 4, p. 25, box 1, Howze Papers (hereafter Howze Oral History); Lyman Lemnitzer oral history, 1972, section 1, p. 4, box 1, Lyman Lemnitzer Papers, USAHEC (hereafter Lemnitzer Oral History).

2. Devore, *When Failure Thrives*, 58–59; Trauschweizer, *Cold War U.S. Army*, 2–3; Department of the Army, Field Manual 100-5, *Field Service Regulations: Operations* (hereafter FM 100-5, *Operations*), September 1954, 6. For contemporary discussions on the differences see Morton H. Halperin, *Limited War in the Nuclear Age* (New York: John Wiley, 1963); Otto Heilbrunn, *Conventional Warfare in the Nuclear Age* (New York: Praeger, 1965); Henry Kissinger, *Nuclear Weapons and Foreign Policy* (Garden City, NY: Doubleday, 1957); and Robert Osgood, *Limited War: The Challenge to American Strategy* (Chicago: University of Chicago Press, 1957).

3. Brian Linn, *Real Soldiering: The US Army in the Aftermath of War, 1815–1980* (Lawrence: University Press of Kansas, 2023), 115.

4. Bernard Brodie, ed., *The Absolute Weapon* (New York: Harcourt Brace, 1946), 74.

5. NSC 68, April 4, 1950, *FRUS: 1950*, vol. 1, 237–39 (Washington, DC: GPO, 1977); Report to the President Pursuant to the President's Directive of January 31, 1950, April 7, 1950, box 13, National Security Council File, Papers of Harry S. Truman, HSTL; Michael J. Hogan, *A Cross of Iron: Harry S. Truman and the Origins of the National Security State, 1945–1954* (Cambridge: Cambridge University Press, 2000), 267; David T. Fautua, "The 'Long Pull' Army: NSC 68, the Korean War, and the Creation of the Cold War U.S. Army," *Journal of Military History* 61, no. 1 (January 1997): 93, 95–96, https://doi.org/10.2307/2953916.

6. Orders, December 31, 1945, box 1, Ridgway Papers; Ridgway and Martin, *Soldier*, 163–74; "Delays imposed by the USSR in the Work of the Military Staff Committee," memorandum, February 3, 1947, box 65, Ridgway Papers; MSC Report No. 8, February 3, 1947, box 8, Ridgway Papers; Jonathan Soffer, *General Matthew B. Ridgway: From Progressivism to Reaganism, 1895–1993* (Westport, CT: Praeger, 1998), 175.

7. Linn, *Elvis's Army*, 26; Taylor, *Swords and Plowshares*, 110, 124–27, 130; Maxwell D. Taylor, "West Point Looks Ahead" (March 1946), Maxwell D. Taylor Letters, USMAA; *Annual Reports of the Superintendent, United States Military Academy, 1947–1989*, Digital Collections, USMAA, see specifically 1947 report, p. 39, https://digital-library.usma.edu/digital/collection/superep/id/53; John M. Taylor, *General Maxwell Taylor*, 92–93, 148; Ingo Trauschweizer, "Berlin Commander: Maxwell Taylor at the Cold War's Frontlines, 1949-51," *Cold War History* 21, no. 1, 52–53.

8. Gavin, *War and Peace*, 109; Gavin, *Airborne Warfare*, 170, 175.

9. James Forrestal, memo, "Establishment of Weapons Systems Evaluation Group," December 11, 1948, box 84, John H. Ohly Papers, HSTL; "Weapons' Values to Be Appraised," *Spokane Daily Chronicle*, December 15, 1948; David C. Elliot, "Project Vista and Nuclear Weapons in Europe," *International Security* 11, no. 1 (1986): 163–83, https://doi.org/10.2307/2538879; OSD to CG Sandia Base, message, May 19, 1949, and DA Office of the Adjutant General, Orders to Maj. Gen. James M. Gavin, April 29, 1949, box 2, entry UD 54-A, Ass Sec Def (R+D) WSEG Research & Records Section, General Decimal File, 1948–53, RG 330, NARA; Brief of Final Report, Project VISTA, September 6, 1951, box 20, entry UD 54-A, Ass Sec Def (R+D) WSEG Research & Records Section, General Decimal File, 1948–53, RG 330, NARA; Gavin Oral History, section 2, pp. 33–34; Gavin, *War and Peace*, 132–37.

10. Ridgway and Martin, *Soldier*, 187, 190; Clay Blair, *The Forgotten War* (New York: Times Books, 1987), 19; Curatola, *Bigger Bombs*, 71; Jeffrey Barlow, *Revolt of the Admirals: The Fight for Naval Aviation, 1945–1950* (Washington, DC: Naval Historical Center, 1993), 184–88.

11. Hanson W. Baldwin, "Need of Training Revealed in Korea," *New York Times*, November 3, 1950; Ridgway and Martin, *Soldier*, 191; *Virginia Pilot and Ledger-Star*, February 12, 1984, box 34, Ridgway Papers; Matthew B. Ridgway, "Man: The Vital Weapon," *Army Combat Forces Journal* 5 (March 1955): 16–19; Thomas Ricks, *The Generals: American Military Command from World War II to Today* (New York: Penguin Books, 2012), 189–90; Gertrude Samuels, "Ridgway—Three Views of a Soldier," *New York Times Magazine*, April 22, 1951, 10; Blair, *Forgotten War*, 570–74; Bradley to MacArthur, message, JCS 88180, April 11,

1951, and Ridgway to JCS, message, C 60965, April 1951, box 14, Harry S. Truman Papers, Korean War File, HSTL; Ridgway and Martin, *Soldier*, 277; CINCFE to JCS, flash message, July 21, 1951, box 115, Harry S. Truman Papers, President's Secretary's Files, General File 1940–1953, HSTL.

12. For reports from Gavin's trip to the Far East with a delegation led by Dr. Edward Bowles see folder: Far East Reports, box 6, entry UD 54-A, Ass Sec Def (R+D) Weapons Systems Evaluation Group Research & Records Section, General Decimal File, 1948–53, RG 330, NARA; James M. Gavin, "The Tactical Use of the Atomic Bomb," *Combat Forces Journal* 1 (November 1950): 9–11; James M. Gavin, "Cavalry, and I Don't Mean Horses," *Harper's*, April 1954.

13. Soffer, *General Matthew B. Ridgway*, 157; Matthew B. Ridgway, "How Europe's Defenses Look to Me," *Saturday Evening Post*, October 10, 1953; Walter Mills, "General Greunther's Headaches," *Collier's*, July 11, 1953; Mitchell, *Matthew B. Ridgway*, 115–17, 123; Ridgway to Bradley, June 19, 1952, box 24, Ridgway Papers; Omar N. Bradley and Clay Blair, *A General's Life* (New York: Simon & Schuster, 1983), 639.

14. Special Orders No. 99, December 8, 1952, box 20, folder 1, Gavin Papers; Booth and Spencer, *Paratrooper*, 341–42; Donald A. Carter, *Forging the Shield: The U.S. Army in Europe, 1951–1952* (Washington, DC: CMH, 2015), 91–94.

15. Taylor, *Swords and Plowshares*, 147–57; Ingo Trauschweizer, *Maxwell Taylor's Cold War: From Berlin to Vietnam* (Lexington: University Press of Kentucky, 2019), 62.

16. Taylor, *Swords and Plowshares*, 152; Jenkins to Taylor, October 6, 1953, Security Classified Correspondence, 1948–1954, 1953, box 411, Korea-10-1953, RG 319, NARA.

17. NSC 162/2, "Basic National Security Policy," October 30, 1953, RG 273, Records of the National Security Council, NARA; Dwight David Eisenhower to Charles E. Wilson, October 20, 1951, box 24, Senate Republican Memo, March 10, 1955, "National Defense under the Republican Administration," box 51, Charles E. Wilson Papers, NLAU (hereafter Wilson Papers); *Annual Report of the Secretary of Defense*, July 1, 1959, to June 30, 1960 (Washington, DC, 1961), 34; "1956 Defense Budget," *Army–Navy–Air Force Register*, January 22, 1955; House Appropriations Subcommittee, *Hearings on Army Appropriations for FY 1955*, 83rd Congress, 2nd session, 1953, 2, 9, 58; Dwight D. Eisenhower, *Mandate for Change* (New York: Doubleday, 1963), 446–47.

18. Robert F. Williams, "Integrating Army Capabilities into Deterrence: The Early Cold War," *Parameters* 53, no. 5 (Winter 2023): 69–82, https://doi.org/10.55540/0031-1723.3260; see also Kenneth A. Osgood, *Total Cold War: Eisenhower's Secret Propaganda Battle at Home and Abroad* (Lawrence: University Press of Kansas, 2006).

19. Linn, *Real Soldiering*, 117; see also Linn, *Echo of Battle: The Army's Way of War* (Cambridge, MA: Harvard University Press, 2007), 156–61; Ridgway and Martin, *Soldier*, 191, 277, 298, 312; Andrew J. Bacevich, "The Paradox of Professionalism: Eisenhower, Ridgway, and the Challenge to Civilian Control, 1953–1955," *Journal of Military History* 61, no. 2 (April 1997): 311–14, 316.

20. John M. Taylor, *General Maxwell Taylor*, 204; Charles L. Bolte, "Do We Need an Army?," *Army Information Digest* 5, no. 6 (June 1954): 3–6. For more on

Project Solarium see "Project Solarium," box 9-10, NSC Series, Subject Subseries, White House Office, Office of the Special Assistant for National Security Affairs: Records 1952-1961, DDEL; George F. Kennan, *Realities of American Foreign Policy* (Princeton, NJ: Princeton University Press, 1954), 84; John Lewis Gaddis, *The Cold War: A New History* (New York: Penguin Books, 2007), 63; Hogan, *Cross of Iron*, 100. For more on the air-atomic strategy and its origins see Michael S. Sherry, *The Rise of American Air Power: The Creation of Armageddon* (New Haven, CT: Yale University Press, 1987).

21. Dwight D. Eisenhower, Annual Message to the Congress on the State of the Union, January 7, 1954, DDEL; Robert H. Ferrell, ed., *The Diary of James C. Hagerty: Eisenhower in Mid-course, 1954-1955* (Bloomington: Indiana University Press, 1983), 182 (Tuesday, February 1, 1955); Joe Buccino, "Ike vs. Ridgway: Lessons for Today from the Philosophical Battle between Two of America's Greatest Military Leaders," *Modern War Institute*, April 14, 2020, https://mwi. usma.edu/ike-vs-ridgway-lessons-today-philosophical-battle-two-americas-greatest-military-leaders/.

22. Jeffrey H. Michaels, "Managing Global Counterinsurgency: The Special Group (CI) 1962-1966," *Journal of Strategic Studies* 35, no. 1 (February 2012): 36-38, https://doi.org/10.1080/01402390.2011.592002.

23. Ridgway and Martin, *Soldier*, 269-70; Matthew B. Ridgway interview with Col. John M. Blair, section 4, p. 31, box 89B, Matthew B. Ridgway papers, USAHEC, Carlisle, PA; Charles E. Wilson, confirmation hearing, January 15, 1953, nomination hearings before the Committee on Armed Services, United States Senate, 83rd Congress, 1st Session, p. 26; Justin Hyde, "Recalling an Awkward Phrase," *Detroit Free Press*, September 14, 2008; Matthew Ridgway interview with Forrest C. Pogue, 1959, p. 38, Pogue interviews, GCMRL; Barksdale Hamlett oral history interview, 1976, Barksdale Hamlett papers, USAHEC; Ridgway and Martin, *Soldier*, 288; House Army Appropriations Subcommittee, *Hearings for FY 1955*, 83rd Congress, 2nd Session, 1954, pp. 45, 49.

24. Hanson W. Baldwin, "Ridgway to the Rescue," *New York Times*, November 23, 1954; Rebecca Grant, "Dien Bien Phu," *Air Force Magazine* 87, no. 8 (August 2004), https://www.airandspaceforces.com/article/0804dien/; John Prados, *Vietnam: The History of an Unwinnable War, 1945-1975* (Lawrence: University Press of Kansas, 2009), 26-31; George C. Herring and Richard H. Immerman, "Eisenhower, Dulles, and Dienbienphu: 'The Day We Didn't Go to War' Revisited," *Journal of American History* 71, no. 2 (September 1984): 343-63, https://doi.org/10.2307/1901759.

25. "Army Outline for U.S. Support to France in Vietnam (Proposed)," memorandum, April 19, 1954, box 19, Gavin Papers; Ridgway to the Secretary of the Army, April 24, 1954, Records of the Chief of Staff of the Army, 1954, RG 319, NARA; James M. Gavin, *Crisis Now* (New York: Random House, 1968), 46-47.

26. US Senate, *Defense Appropriations Subcommittee, Defense Appropriations Hearings for FY 1956*, April 4-June 6, 1955, p. 105; Hanson W. Baldwin, "Ridgway vs. Eisenhower: A Review of the Apparent Contradiction in Their Remarks on Manpower Slashes," *New York Times*, January 24, 1956; Ridgway quoted in "1956 Defense Budget," *Army-Navy-Air Force Register*, January 22, 1955.

27. "Army to Retain Gen. Ridgway," *Army–Navy–Air Force Register*, January 22, 1955; "President Pays Honor to Retiring General Ridgway," *New York Times*, June 29, 1955; Ridgway and Martin, *Soldier*, 260; in Ferrell, *Diary of James C. Hagerty*, 182 (Tuesday, February 1, 1955); Ridgway to Wilson, June 27, 1955, box 18, Ridgway Papers; "Ridgway Challenges President on Troops," *New York Times*, January 17, 1956.

28. Gavin, *War and Peace*, 155; Matthew B. Ridgway, *Saturday Evening Post*, January 21 to February 25, 1956; Soffer, *General Matthew B. Ridgway*, 186; Bacevich, "Paradox of Professionalism," 307, 311, 332–33; Conrad Crane, "Matthew Ridgway and the Value of Persistent Dissent," *Parameters* 51, no. 2 (Summer 2021): 18, https://doi.org/10.55540/0031-1723.3064.

29. Taylor, *Swords and Plowshares*, 152; Trauschweizer, *Maxwell Taylor's Cold War*, 67–68.

30. Palmer reinjured an old horse-inflicted injury while attempting to qualify as a parachutist in 1950: see Maj. Gen. Williston B. Palmer to Maxwell D. Taylor, May 29, 1959, box 5, Williston B. Palmer Papers, USMAA (hereafter Palmer Papers, USMAA); Palmer was quoting Brucker in a memorandum: see Vice Chief of Staff Memorandum, "Good Will for the Army," August 17, 1955, box 5, Palmer Papers, USMAA; Maxwell D. Taylor, Oral History interview, section 3, pp. 32–33, 1974, Maxwell Taylor Papers, USAHEC.

31. Office of the Adjutant General, "Army Position on Major Issues," memorandum, April 4, 1956, box 9, Palmer Papers, USMAA; Ricks, *Generals*, 219; Address by General Maxwell D. Taylor, Chief of Staff, United States Army, at the First Annual Meeting of the Association of the United States Army, Fort Benning, GA, October 22, 1955, Taylor Papers, Special Collections, NDU; Chief of Staff, US Army, to the Joint Chiefs, memorandum, "Army, Naval, and Air Force Forces under Imposed Budgetary Limitations," July 9, 1956, and Chief of Staff, July 25, 1957, statement at National Security Council meeting, box 6, White House Office, Office of the Staff Secretary, 1952-61 Subject Series, DoD Subseries, DDEL; Taylor, *Swords and Plowshares*, 158.

32. Andrew Birtle, *US Army Counterinsurgency and Contingency Operations Doctrine, 1942–1976* (Washington, DC: CMH, 2006), 157–59, 179n59; W. W. Culp, "Resident Courses of Instruction," *Military Review* 36 (May 1956): 17, 20.

33. Gavin, "Cavalry, and I don't Mean Horses"; Gavin quoted in Booth and Spencer, *Paratrooper*, 352; Gavin, *War and Peace*, 124; James M. Gavin, "Arms Vigilance for Peace," *Ordnance* 39, no. 209 (March–April 1955): 716–19.

34. Taylor oral history, section 3, p. 31; Maxwell D. Taylor, "The Army—an Appeal to the Record," speech to the Al Smith Memorial Dinner, New York, October 20, 1955, box 5, Taylor Papers, NDU; transcript of press conference with Maxwell D. Taylor, January 10, 1956, box 37, Administrative Series, Dwight D. Eisenhower Papers as President (Ann Whitman file), DDEL.

35. Anthony Leviero, "Military Forces Split by Conflict on Arms Policies," *New York Times*, May 19, 1956; James E. Hewes, *From Root to McNamara: Army Organization and Administration, 1900–1963* (Washington, DC: CMH, 1975), 239-24; see "Employment Assignment of the IRBM," box 9, Palmer Papers, USMAA, and "A U.S. Military Program," June 1961, Presidential Papers, National Security Files, Series 4: Departments and Agencies, box 269, JFKL.

36. Anthony Leviero, "Radford Seeking 800,000 Man Cut; 3 Services Resist," *New York Times*, July 13, 1956; Gavin quoted in Booth and Spencer, *Paratrooper*, 356; Correspondence with James Hollingsworth and Lyal Metheny, 1977–1978, box 20, folder 4, Gavin Papers; Douglas Kinnard, "Civil-Military Relations: The President and the General," *Parameters* 15, no. 1 (1985): 19–29, https://doi.org/10.55540/0031-1723.1387; Trauschweizer, *Maxwell Taylor's Cold War*, 79; David Halberstam, *The Best and the Brightest* (New York: Random House, 1969), 473–77.

37. US Senate, *Study of Airpower*, hearings before the Subcommittee on the Armed Services, United States Senate, 84th Congress, 2nd Session, April 16–June 1, 1956, vol. 1, 860–62; Anthony Leviero, "Army Fails to Bar Bomb Testimony," *New York Times*, June 29, 1956; Linda McFarland, *Cold War Strategist: Stuart Symington and the Search for National Security* (Westport, CT: Praeger, 2001), 78–80.

38. Gavin Oral History, section 2, pp. 40–45; US Senate, *Inquiry into Satellite and Missile Programs, Preparedness Investigating Subcommittee, Senate Armed Services Committee*, 85th Congress, 1st Session, December 13, 1957, 505–10; John Norris, "Gen. Gavin Would End Joint Staff," *Washington Post and Times Herald*, December 14, 1957; Memo to Adjutant General, subj. "Voluntary Retirement," December 23, 1957, box 20, folder 13, Gavin Papers.

39. Gavin quoted in Hanson Baldwin, "Gavin Explains Decision; Warns of 'Another Korea,'" *New York Times*, January 12, 1958; see also folder 273: Correspondence with James M. Gavin, box 6, Hanson Weightman Baldwin Papers (MS 54), Manuscripts and Archives, Yale University Library (hereafter Baldwin Papers); Alvin Schuster, "Gen. Gavin, Missile Aide, to Quit; Criticized Joint Chiefs System," *New York Times*, January 5, 1958; Allen Drury, "Gavin's Quitting Will Be Studied by Senate Unit," *New York Times*, January 6, 1958; "Gavin Urged as Senator," *New York Times*, January 8, 1958; James Reston, "President Asks More Missiles, Further Aid, Pentagon Unit," *New York Times*, January 10, 1958; "Gavin Retires, Backs Atomic Tests," *New York Times*, April 1, 1958.

40. Ridgway and Martin, *Soldier*; Gavin, *War and Peace*; Maxwell Taylor, *The Uncertain Trumpet* (New York: Harper & Bros., 1959); National Security Council Report, "NSC 5906/1, Basic National Security Policy," August 5, 1959, *FRUS, 1958–1960*, vol. 3, National Security Policy, Arms Control and Disarmament, Document 70; Raymond Millen, "The Post-Korean War Drawdown under the Eisenhower Administration," in *Drawdown: The American Way of Postwar*, ed. Jason W. Warren (New York: NYU Press), 201–2. For a view from the field-grade level on agreeing with scholarly criticism of the New Look see G. A. Lincoln and Amos A. Jordan Jr., "Limited War and the Scholars," *Military Review* 37, no. 10 (January 1958): 50–60.

41. Statement of General Maxwell D. Taylor, Chief of Staff, U.S. Army, before the Subcommittee on Department of Defense Appropriations, House of Representatives, Relative to the Department of the Army Budget for Fiscal Year 1957, Taylor Papers, NDU; "Security through Deterrence," address by Gen. Maxwell D. Taylor before the Army War College, Carlisle Barracks, PA, August 20, 1956, Selected Speeches as Army Chief of Staff, 1955–1959, box 4, Taylor Papers, NDU.

42. John F. Kennedy, *The Strategy of Peace*, ed. Allan Nevins (New York: Harper & Row, 1960), 34. See also Christopher A. Preble, *John F. Kennedy and the Missile Gap* (DeKalb: Northern Illinois University Press, 2004); "Text of Kennedy's Speech to Senate Advocating New Approach on Foreign Policy," *New York Times*, June 5, 1960; Martin Clemis, "Once Again with the High and Mighty," in Warren, *Drawdown*, 213; Kennedy, *Strategy of Peace*, 184; James N. Giglio, *The Presidency of John F. Kennedy* (Lawrence: University Press of Kansas, 2006), 48.

43. Kennedy to Taylor, letter, n.d., Papers of John F. Kennedy, Presidential Papers, President's Office Files, Special Correspondence, Taylor, General Maxwell D., JFKL; Trauschweizer, *Maxwell Taylor's Cold War*, 101–36.

44. Maxwell Taylor oral history interview by Elspeth Rostow, April 12, 1964, 1–2, John F. Kennedy Oral History Collection, JFKL; Correspondence between Cornelius Ryan and James Gavin, 1961, box 2, Supplementary Material, Cornelius Ryan Collection of World War II Papers, Mahn Center for Archives and Special Collections, Athens, OH; Booth and Spencer, *Paratrooper*, 398–418.

45. Kennedy, *Strategy of Peace*, 37–38; Dealie, *Always at War*, 200; Taylor, *Uncertain Trumpet*, 6; "A U.S. Military Program," June 1961, Presidential Papers, National Security Files, Series 4: Departments and Agencies, box 269, JFKL; FM 100–5, *Operations*, 1962; Jane E. Stromseth, *The Origins of Flexible Response: NATO's Debate over Strategy in the 1960s* (London: Macmillan, 1988), 1.

46. Kennedy, *Strategy of Peace*, 33–45; Giglio, *Presidency of John F. Kennedy*, 47–49; John Lewis Gaddis, *Strategies of Containment: A Critical Appraisal of American National Security Policy during the Cold War* (New York: Oxford University Press, 2005), 212–14, 217; Preble, *John F. Kennedy and the Missile Gap*, 4, 9; Kaplan, *To Kill Nations*, 174.

47. Stromseth, *Origins of Flexible Response*, 26; Taylor, *Uncertain Trumpet*, 159; Robert S. McNamara, *The Essence of Security: Reflections in Office* (New York: Harper & Row, 1968), 82; McNamara to Kennedy, May 10, 1961, *FRUS, 1961–1963*, vol. 8, National Security Policy, Document 27; Clemis, "Once Again with the High and Mighty," 216; Department of Defense, *Annual Report for Fiscal Year 1964* (Washington, DC: GPO, 1965), 16–17; "Text of the President's Message and Budget Analysis," *New York Times*, January 19, 1962.

48. Maxwell D. Taylor, "The American Soldier," remarks for commencement address, West Point, June 5, 1963, Maxwell Taylor Letters, USMAA; Kennedy, Special Message to the Congress on the Defense Budget, March 28, 1961, Document 99, JFK PP: 1961, 230–31.

49. Conversations between President Kennedy and Chairman Khrushchev, June 3–4, 1961, *FRUS, 1961–1963*, vol. 5, Soviet Union, Document 95; Michaels, "Managing Global Counterinsurgency," 38–42; NSAM 124 in *FRUS: 1961–1963*, vol. 2, Document 26.

50. McNamara to Kennedy, "Report to the President, FY 1961 and 1962 military programs and budgets," Annex A, attachment 2: Limited War Proposals, February 21, 1961, Papers of John F. Kennedy, Presidential Papers, President's Office Files, Departments and Agencies: Defense, JFKL; Gaddis, *Strategies of Containment*, 216; Boyd L. Dastrup, *The US Army Command and General Staff College: A Centennial History* (Manhattan, KS: Sunflower University Press, 1982), 110–11.

51. Andrew F. Krepinevich, *The Army and Vietnam* (Baltimore: Johns Hopkins University Press, 1986), 108–9; Kenneth Finlayson, "Lieutenant General William P. Yarborough," *Veritas* 2, no. 2 (2006): 45, https://arsof-history.org/articles/pdf/v2n2_yarborough.pdf.

52. Walt W. Rostow, draft, "Basic National Security Policy," March 26, 1962, 110–11, in *FRUS: 1961–1963*, vol. 8, Document 70; Gaddis, *Strategies of Containment*, 223; Michael Latham, *Modernization as Ideology: American Social Science and "Nation Building" in the Kennedy Era* (Chapel Hill: University of North Carolina Press, 2000); David Steigerwald, *The Sixties and the End of Modern America* (New York: St. Martin's, 1995), 13. For more from Rostow see W. W. Rostow, *The Diffusion of Power, 1957–1972* (New York: Macmillan, 1972), 216, 424–25, 429–30. For more on the genesis of the Peace Corps see Gavin correspondence, Peace Corps, folder 3, box 29, Gavin Papers. See also Coates Redmon, *Come as You Are: The Peace Corps Story* (New York: Harcourt Brave Jovanovich, 1986), 3–21, and Booth and Spencer, *Paratrooper*, 399–401; Kevin Boyle, *The UAW and the Heyday of American Liberalism, 1945–1968* (Ithaca, NY: Cornell University Press, 1995), 144; Hubert Humphrey, *The Education of a Public Man* (New York: Doubleday, 1976), 184.

53. Trauschweizer, *Maxwell Taylor's Cold War*, 103; Sepp, "Pentomic Puzzle," 11; Lewis Sorley, *Thunderbolt: General Creighton Abrams and the Army of His Times* (New York: Simon & Schuster, 1992); Adrian Lewis, *The American Culture of War: The History of US Military Force from World War II to Operation Iraqi Freedom* (New York: Routledge, 2007), 220.

54. John J. Tolson, "The Future of Army Aviation," Army War College student paper (April 1, 1953), 28–30, USAHEC; Taylor Oral History, section 3, p. 31, USAHEC.

55. Brian McAllister Linn, "Eisenhower, the Army, and the American Way of War," Lecture, Eisenhower Lecture Series at Kansas State University, 2003; Norman E. Martin, "Dien Bien Phu and the Future of Airborne Operations," *Military Review* 36, no. 3 (June 1956): 19; Bruce C. Clarke, "Abe," *Armor* 84, no. 1 (January-February 1975): 17.

56. National Security Council Report, "NSC 5906/1," August 5, 1959, *FRUS: 1958–1960*, National Security Policy, Arms Control and Disarmament, vol. 3, Document 70; Department of Defense Reorganization Act of 1958, August 6, 1958, in *The Department of Defense: Documents on Establishment and Organization, 1947–1978*, ed. Alice C. Cole (Washington, DC: Office of the Secretary of Defense Historical Office, 1978), 188–230; Donald A. Carter, "Eisenhower versus the Generals," *Journal of Military History* 71, no. 4 (October 2007): 1198–99, https://doi.org/10.1353/jmh.2007.a222498; Hewes, *From Root to McNamara*, 297.

57. Halberstam, *Best and the Brightest*, 163; Gavin Oral History, 1975, pp. 43–45, Gavin Papers.

4. The Airborne Influence on Atomic Warfare

1. US House of Representatives, *Committee on Armed Services, Hearings before Committee on Armed Services*, 85th Cong., 1st Session, 1956, 143–44; Theodore

C. Mataxis and Seymour L. Goldberg, *Nuclear Tactics: Weapons and Firepower in the Pentomic Division, Battle Group, and Company* (Harrisburg, PA: Military Service, 1958), 1.

2. "Your Army in the Atomic Age," remarks of Secretary of the Army Frank Pace Jr. before the National Wool Manufacturers Association Convention, New York City, May 8, 1952, box 15, Frank Pace Jr. Papers, HSTL; Kaplan, *To Kill Nations*, 2; "War and Peace in the Nuclear Age; Bigger Bang for the Buck; an Interview with James Gavin, 1986," February 25, 1986, GBH Archives, Boston; "Major Duties and Responsibilities of the Secretary of the Army," September 21, 1955, Briefing for *Meet the Press* appearance 1955, box 16, Wilbur M. Brucker Papers, BHL; Lawrence Freedman and Jeffrey Michaels, *The Evolution of Nuclear Strategy*, 4th ed. (London: Palgrave Macmillan, 2019), 135; President Dwight Eisenhower quoted in "1956 Defense Budget," *Army–Navy–Air Force Register*, January 22, 1955; Gavin quote from Linn, *Elvis's Army*, 73.

3. Taylor, *Swords and Plowshares*, 171; Trauschweizer, *Maxwell Taylor's Cold War*, 66; Linn, *Elvis's Army*, 74.

4. Mataxis and Goldberg, *Nuclear Tactics*, 4; Dr. Ellis A. Johnson, staff memorandum, "The Long Range Future of the US Army," July 11, 1955, Operations Research Office, Johns Hopkins University, box 6, DoD Subseries, 1952–61 Subject Series, White House Office of the Staff Secretary, DDEL; Gavin, *Airborne*, 170, 181–83; Gavin, *War and Peace*, 112; Gavin, "Tactical Use of the Atomic Bomb," 11; Linn, *Real Soldiering*, 129; "Gavin Says Air Mechanization Will Win Wars of the Future," *Army and Navy Bulletin* 3 (April 5, 1947): 3.

5. Elliot, "Project Vista," 164; Gavin, *War and Peace*, 129–35.

6. Theodore H. White, "An Interview with General Gavin . . . Tomorrow's Battlefield," *Army Combat Forces Journal* 5 (March 1955): 22; Gavin, *War and Peace*, 137.

7. White, "Interview with General Gavin," 23; Gavin, "Tactical Use of the Atomic Bomb," 11; Gavin, *War and Peace*, 137–39; "New Divisional Organization," *Army–Navy–Air Force Journal and Register*, February 12, 1955, 1–2.

8. Office of the Chief of Army Field Forces, memorandum, Subj: Tactical Employment of the Atomic Bomb, October 7, 1951, file 00.9/35, box 4, entry 55B, RG 337, NARA; Department of the Army, *Field Service Regulations: Operations*, Field Manual 100-5 (hereafter DA, FM 100-5, *Operations*), 1954, 40; Frank Pace Jr., "Your Army in the Atomic Age"; Maxwell Taylor, "Missions of the United States Army," remarks at Armed Forces Staff College, November 1956, Taylor Papers, NDU; Maxwell Taylor, "Security through Balanced Deterrence," Calvin Bullock Forum, December 10, 1956, Taylor Papers, NDU; "Gavin Says Air Mechanization Will Win Wars of the Future," *Army and Navy Bulletin* 3 (April 5, 1947): 3.

9. Harold H. Martin, "Paratrooper in the Pentagon," *Saturday Evening Post*, August 28, 1954, 81; Paul Disney, "Armor in Atomic Warfare," *Armor* 63 (May–June 1953): 30–31; Paul D. Adams, "Final Report, Exercise Sage Brush," January 1956, p. 3, Reference Material box 5, Training Exercises, WMM82; Linn, *Echo of Battle*, 175.

10. George C. Reinhardt and William R. Kintner, *Atomic Weapons in Land Combat* (Harrisburg, PA: Military Service, 1953), 49–50; Mataxis and Goldberg,

Nuclear Tactics, 164, 211; Hanson W. Baldwin, "Infantry Adjusts Role," *New York Times*, November 23, 1953; Bacevich, *Pentomic Era*, 109–10.

11. DA, FM, *Operations*, 1954, 96; Mataxis and Goldberg, *Nuclear Tactics*, 211.

12. Reinhardt and Kintner, *Atomic Weapons in Land Combat*, 51–52; DA, FM 100–5, *Operations*, 1954, 117; Mataxis and Goldberg, *Nuclear Tactics*, 127–29, 131; Willard G. Wyman, "Let's Get Going on Our New Combinations for Combat," *Army* 6 (July 1956): 40.

13. Hamilton H. Howze, "Combat Tactics for Tomorrow's Army," *Army* 8 (October 1958): 28; Frank W. Moorman, "Logistical Problems in Future Warfare," *Military Review*, July 1950, 9; White, "Interview with General Gavin," 21.

14. Donald A. Carter, "War Games in Europe: The U.S. Army Experiments with Atomic Doctrine," in *Blueprints for Battle: Planning for War in Central Europe, 1948–1968*, ed. Jan Hoffenaar and Dieter Krüger (Lexington: University Press of Kentucky, 2012), 135; Trauschweizer, *Cold War U.S. Army*, 57; Review of Initial Manuscript, C3 to FM 100–5, Field Service Regulations–Operations, February 15, 1956, box 17, and Col. Adam S. Buyonski to CG, CONARC, December 20, 1958, Final Manuscript of FM 100–5, FSR, Operations, box 30, entry UD 3, CGSC, Fort Leavenworth, RG 546 Records of the United States Continental Army Command, NARA.

15. Script of Interview of Chief of Staff by Richard Harkness, October 13, 1954, box 77, Ridgway Papers; Ridgway and Martin, *Soldier*, 298; Trauschweizer, *Cold War U.S. Army*, 57; Bacevich, *Pentomic Era*, 108–10.

16. *Semiannual Report of the Secretary of the Army, FY 1958*, p. 106, box 784, folder 319.1, Office of the Secretary of the Army, General Correspondence January 1957–December 1960, RG 335, NARA; Infantry School Comments to Report of Army Tests Exercise Sage Brush, March 21, 1956, box 3, entry A1 109, Security Classified General Records, Infantry School, R and D General Correspondence, RG 546, NARA; General Plan, Exercise Flash Burn, April–May 1954, box 57, entry UD 35176: General Records [XVIII (Airborne) Corps, 1951–1963], RG 338, NARA.

17. Maxwell D. Taylor, "The Army, Today and Tomorrow," speech to AUSA annual convention, Washington, DC, October 25, 1956, box 5, Taylor Papers, NDU; Taylor, *Swords and Plowshares*, 152–53; "The U.S. Army Today," n.d. (possibly 1958), folder: Miscellaneous papers of Secretary of the Army, box 17, Brucker Papers, BHL; Henry E. Kelly, "Verbal Defense," *Military Review* 35 (October 1955), 48, 51–52; Francis X. Bradley, "The Fallacy of Dual Capability," *Army* 10 (October 1959): 18–19, and Arthur S. Collins, "The Other Side of the Atom," *Army* 10 (November 1959): 18–19; William E. DePuy, "The Case for Dual Capability," *Army* 10 (January 1960), 32–34, 38.

18. Douglas Lindsey, "No Time for Despair," *Armor* 65 (May–June 1956), 38–39; Jane Erikson, "US Surgeon Lindsey, Korean War Vet, Dies," *Arizona Daily Star*, January 5, 2006; Nevada Test Organization, press release, August 24, 1957, RM, Training Exercises, box 6, WMM82; "Troops Carry on after Atomic Blast 2.7 Miles Away," *New York Times*, September 3, 1957; Paul D. Adams, "Final Report, Exercise Sage Brush," January 1956, p. 2, Reference Material, box 5, Training Exercises, WMM82.

19. Michael Evans, "The Primacy of Doctrine: The United States Army and Military Innovation and Reform, 1945–1995," Army Occasional Paper No. 1, Washington, DC, 1996, 5.

20. Key West Agreement, "Functions of the Armed Forces and the Joint Chiefs of Staff," (revision) October 1, 1953, original dated April 21, 1948, appendix A, Association of the United States Army, "The Security of the Nation: A Study of Current Problems of National Defense" (Washington, DC, 1957), p. 21, box 24, Wilson Papers; Kenneth Campbell, "Pace Urges Army Built on Science," *New York Times*, June 7, 1950; Taylor, *Swords and Plowshares*, 171.

21. Carter, "War Games in Europe," 138; Linn, *Elvis's Army*, 104; K. D. Nichols, "Atomic Guns," *US News & World Report*, July 10, 1953; remarks of Secretary of the Army Frank Pace Jr. at Public Demonstration of the AT-280 Gun, Aberdeen Proving Ground, October 15, 1952, box 15, Frank Pace Jr. Papers, HSTL; Report on the 280mm gun, ca. 1952, box 16, 280 MM Gun 1952, box 16, J. Lawton Collins Papers, DDEL.

22. Key West Agreement, "Functions of the Armed Forces and the Joint Chiefs of Staff," appendix A, 21; DA, AR 525-50, *Army Air Defense Operations* (Washington, DC, 1956); "Missiles and Space," presentation, Maj. Gen. John B. Medaris, September 25, 1959, Sixth Annual Conference of Civilian Aides to the Secretary of the Army, September 23–25, 1959, folder: Papers, 1959, box 17, Brucker Papers; Gavin Oral History, section 2, p. 35; Richard E. Mooney, "Army's Explorer Cheers Congress," *New York Times*, February 2, 1958.

23. Key West Agreement, "Functions of the Armed Forces and the Joint Chiefs of Staff," appendix A, 21; Charles E. Wilson, memorandum, "Clarification of Roles and Missions to Improve the Effectiveness of Operation of the Department of Defense," November 26, 1956, box 6, DoD Subseries, 1952–61, White House Office of the Staff Secretary, DDEL; Anthony Leviero, "Air Force Calls Army Nike Unfit to Guard Nation," *New York Times*, May 21, 1956; Anthony Leviero, "Air Force Doubts Carriers' Value," *New York Times*, May 20, 1956; "The Army's Bird in Hand," *Air Force Magazine*, June 1956, 42; Department of the Army, *Army Air Defense Operations*, AR 525-50 (Washington, DC, 1956).

24. Gavin, *War and Peace*, 161, 218, 226; Linn, *Elvis's Army*, 105; Maxwell Taylor, remarks to Army Commander's Conference, Fort Bliss, TX, April 5, 1956, box 9, Taylor Papers, NDU.

25. Gavin, *Airborne Warfare*, 172; Gavin, *War and Peace*, 112–13; Sidney Shalett, "Arnold Reveals Secret Weapons, Bomber Surpassing All Others," *New York Times*, August 18, 1945.

26. Department of the Army, *Army Missiles*, AR 525-20 (Washington, DC, 1956), 1; James W. Bragg, *Development of the Corporal: The Embryo of the Army Missile Program* (Redstone Arsenal, AL: Army Ballistic Missile Agency, 1961), 115; Gavin, *War and Peace*, 145, 154; Department of the Army, *Army Missiles and Rockets*, DA Pamphlet 355-13, May 1958, 36.

27. Gavin, *War and Peace*, 145, 153; Final Report of the Director of Guided Missiles, Office of the Secretary of Defense, September 17, 1953, folder May 1956 (1), box 7, Correspondence, May 1956–June 1957, Brucker Papers; "Employment Assignment of the IRBM," box 9, Palmer Papers, USMAA; DA

Pamphlet 355-13, May 1958, 8; William M. Blair, "Wilson Restricts Army on Missiles and Air Program," *New York Times*, November 27, 1956.

28. Gavin, *War and Peace*, 155; US Senate, *Study of Airpower*, Hearings before the Subcommittee on the Air Force of the Committee on Armed Services, 84th Congress, 2nd Session (Washington, DC: GPO, 1956), 1287; Association of the United States Army, "The Security of the Nation: A Study of Current Problems of National Defense," (Washington, DC, 1957), 17, box 24, Wilson Papers; "Missile Unit Set Up," *New York Times*, January 18, 1958; Jack Raymond, "Thor Is Selected Over the Jupiter," *New York Times*, September 29, 1958.

29. Quote in Bacevich, *Pentomic Era*, 86; Mataxis and Goldberg, *Nuclear Tactics*, 32, 39, 36; DA Pamphlet 355-13, May 1958, 12; "Army's New Guided Missile Takes Off," *New York Times*, September 20, 1951; "New Atom Missile Ready for Troops," *New York Times*, April 18, 1954; "Missiles and Space," presentation, Maj. Gen. John B. Medaris, September 25, 1959.

30. Mataxis and Goldberg, *Nuclear Tactics*, 41; DA Pamphlet 355-13, May 1958; Janice E. McKenney, *Organizational History of Field Artillery 1775–2003* (Washington, DC: CMH, 2007), 230-34.

31. Maxwell Taylor, "Military Objectives of the Army, 1960–1962," Address at the Secretaries' Conference, Quantico, VA, June 21, 1958, Taylor Papers, NDU; DA Pamphlet 355-13, May 1958, 17.

32. Leonard C. Weston, *Project Management of the Davy Crockett Weapons System, 1958–1962* (Rock Island, IL: Rock Island Historical Branch, 1964), 12; Matthew Seelinger, "The M28/M29 Davy Crockett Nuclear Weapon System," Army Historical Foundation, https://Armyhistory.org/the-m28m29-davy-crockett-nuclear-weapon-system/; DA, Field Manual 23-20, *Davy Crockett Weapons System in Infantry and Armor Units*, 1961; Linn, *Elvis's Army*, 108.

33. US Senate, *Inquiry into Satellite and Missile Programs*, 490; *Semiannual Report of the Secretary of the Army January 1–June 30, 1957* (Washington, DC, 1958), 104; Gavin, *War and Peace*, 161.

34. Bacevich, *Pentomic Era*, 103; Trauschweizer, *Cold War U.S. Army*, 81; Maj. Gen. James M. Gavin to CAFF, "Airborne Role of the Infantry Center," August 21, 1954, 358 (Abn) file, box 24, E 55F RG 337, Records of Army Ground Forces, NARA II; William E. Ekman, "The Helicopter in Our Future Army," March 31, 1953, Army War College student paper, USAHEC.

35. Linn, *Elvis's Army*, 114–15; Anthony L. Wermuth, "Modernization-Minus," *Army* 9 (October 1958): 31; Ellsberg quote in Halberstam, *Best and the Brightest*, 22.

36. Taylor, remarks at the Army War College, Carlisle Barracks, PA, August 20, 1956, box 9, Taylor Papers, NDU; Al Gruenther to Gen. Palmer, July 3, 1958, box 2, Palmer Papers, USMAA.

37. Meeting of Subcommittee III, study group on Nuclear Weapons and Foreign Policy, February 15, 1956, vol. 61, Council on Foreign Relations, box 20, Gavin Papers; White, "Interview with General Gavin," 22; Gavin, *War and Peace*, 137; *Semiannual Report of the Secretary of the Army, FY 1958*, 106; Bacevich, *Pentomic Era*, 105.

38. "Text of President Eisenhower's Budget Message to Congress for the Fiscal Year 1955," *New York Times*, January 22, 1954; OCoS to OCAFF,

memorandum, April 19, 1954, subj: "Organizational Studies to Improve the Army Combat Potential-to-Manpower Ratio," RG 337, NARA; John B. Wilson, *Maneuver and Firepower: The Evolution of Divisions and Separate Brigades* (Washington, DC: CMH, 1998), 267; Matthew Ridgway, "The Army's Role in National Defense," *Army Information Digest* 9, no. 5 (May 1954): 25.

39. Advanced Study Group, Project Binnacle: Concepts and Doctrine for Future Warfare, Conventional or Nuclear, 1960–1970 (January 1955), USAHEC; Linn, *Elvis's Army*, 83.

40. Linn, *Elvis's Army*, 52; G-3 to CG, CONRAC, November 17, 1954, "Organization of the Army during the Period FY 1960–1970," entry P 218, Records of the Deputy Chief of Staff for Military Operations, Central Decimal Files, 1954, RG 319, NARA; Briefings and Conferences ATFA-1 file, entry 30B, box 488, RG 337, NARA; Wilson, *Maneuver and Firepower*, 267; "New Divisional Organization," *Army–Navy–Air Force Register*, February 12, 1955.

41. Brig. Gen. Carl F. Fritzsche to Commandant, CGSC, memorandum, "Project ATFA-1," July 10, 1954, box 2, entry A1 109, Security Classified General Records, Infantry School, R&D General Correspondence, RG 546, NARA.

42. Paul D. Adams, "Final Report, Exercise Sage Brush," January 1956, 4, Reference Material box 5, Training Exercises, WMM82; Maj. Gen. Joseph H. Harper, "Third Infantry Division's Final Evaluation Report on the ATFA Infantry Division (TOE 7T)," February 15, 1956, box 3, entry A1 109, Security Classified General Records, Infantry School, R and D General Correspondence, RG 546, NARA; Office of the Deputy Maneuver Director, Headquarters, "Exercise Sagebrush (Oct. 31, 1955)," 2, and Lt. Col. Frank Meszar, "Report of Phase V: Exercise Sagebrush," Infantry School Tactical Department (January 10, 1956), 3–5, box 5, entry A1 109, RG 546, NARA; Paul L. Davis, "Organization of the Army," Report of the Advisory Committee on Army Organization, December 18, 1953, 39, box 12, and Williston B. Palmer, memorandum, "The Continental Army Command," May 1, 1957, box 8, Palmer Papers, USMAA.

43. Department of the Army, Deputy Chief of Staff for Military Operations, Briefing for Chief of Staff on Army Organization 1960–70 (PENTANA), May 15, 1956, USAHEC; "Reorganization of the Airborne Division," June 27, 1956, entry P 218, Deputy Chief of Staff for Military Operations, Central Decimal Files, 1956–1964, 1956 records, box 66, RG 319, NARA; John M. Taylor, *General Maxwell Taylor*, 198–99; Linn, *Elvis's Army*, 90–91.

44. Reorganization of the Airborne Division," June 27, 1956, entry P 218, Deputy Chief of Staff for Military Operations, Central Decimal Files, 1956–1964, 1956 records, box 66, RG 319, NARA; "New Divisional Organization," *Army Information Digest* 12, no. 5 (May 1957): 198–99.

45. T. L. Sherburne, "Reorganizing the 101st Airborne Division: An Interim Report," *Army Information Digest* 12, no. 6 (June 1957): 13; Wilson, *Maneuver and Firepower*, 272–76; XVIII Airborne Corps, memorandum, "Review of Training Tests," July 20, 1956, box 7, entry P 50470, 3rd Army, G-3 Records Relating to Training 1957–1960, RG 338, NARA; Detailed Plan of Test, Troop Test JUMP LIGHT, HQ 3rd Army, box 57, entry UD 35176: General Records [XVIII (Airborne) Corps, 1951–1963], RG 338, NARA; "Mission and Use of 101st Airborne Division," memorandum, March 12, 1956, entry P 218, Deputy Chief of

Staff for Military Operations, Central Decimal Files, 1956–1964, 1956 records, box 62, RG 319, NARA.

46. Transcript of press conference with Maxwell D. Taylor, January 10, 1956, box 37, Administrative Series, Dwight D. Eisenhower Papers as President (Ann Whitman file), DDEL; Ernest B. Furgurson, *Westmoreland: The Inevitable General* (Boston: Little, Brown, 1968), 239–40; CDCOPS to CSA, memorandum, May 2, 1957, entry P 218, Deputy Chief of Staff for Military Operations, Central Decimal Files, 1956–1964, 1957 records, RG 319, NARA; Department of Defense, *Semiannual Report of the Secretary of the Army, FY 1958*, 105.

47. Taylor, *Swords and Plowshares*, 171; "New Divisional Organization," *Army Information Digest* 12, no. 5 (May 1957): 17; Paul C. Jussel, "Intimidating the World: The United States Atomic Army, 1956–1960" (PhD diss., Ohio State University, 2004), 167.

48. Sherburne, "Reorganizing the 101st Airborne Division," 15–18; "New Divisional Organization," *Army Information Digest* 12, no. 5 (May 1957): 22; Everett C. Royal, "The Team of Mobile Warfare: Armor and Airborne," *Armor* 65 (March–April 1955): 4.

49. Taylor, *Swords and Plowshares*, 171; Sepp, "Pentomic Puzzle," 9; John J. McGrath, *The Brigade: A History of Its Organization and Employment in the US Army* (Fort Leavenworth, KS: CSI Press, 2004), 59; Taylor Oral History, section 3, p. 36, Taylor Papers, USAHEC; William M. Donnelly, "Bilko's Army: A Crisis in Command?," *Journal of Military History* 75, no. 4 (October 2011): 1183–1215.

50. Trauschweizer, *Cold War U.S. Army*, 96–98; Gavin Oral History, section 2, p. 45; White, "Interview with General Gavin," 21.

51. C. D. Eddleman, "The Pentomic Reorganization—a Status Report," *Army Information Digest* 13, no. 9 (September 1958): 4; "Phased Reorganization Schedules," in Reorganization of Current Infantry Divisions, October 15, 1957, entry P 218, Deputy Chief of Staff for Military Operations, Central Decimal Files, 1956–1964, 1957 records, RG 319, NARA; Anthony L. Wermuth, "Modernization-Minus," *Army* 9, no. 3 (October 1958): 31.

52. Freedman and Michaels, *Evolution of Nuclear Strategy*, 151; "MOMAR I: Modern Mobile Army, 1965–1970," Staff Study, HQ, CONARC, February 10, 1960, 1–4, USAHEC; Lyman L. Lemnitzer, "Why We Need a Modern Army," *Army* 10, no. 2 (September 1959): 16–21; Davis, *Challenge of Adaptation*, 29.

53. Trauschweizer, *Cold War U.S. Army*, 99–106.

54. Trauschweizer, 111–11; Carter, *US Army before Vietnam*, 37.

55. "MOMAR II," US Army CGSC Study, November 16, 1960, USAHEC; Clyde D. Eddleman, Address to 15th Annual Conference of the Aerospace Industries Association, Williamsburg, VA, June 8, 1961, box 1, Eddleman Papers; Wilson, *Maneuver and Firepower*, 305–6; Defense Budget FY '64, vol. 1: Recommended General Purpose Forces, December 4, 1962, Presidential Papers, National Security Files, Department of Defense, Papers of John F. Kennedy, JFKL; Appraisal of Capabilities of Conventional Forces, May 12, 1961, Presidential Papers, President's Office Files, Papers of John F. Kennedy, JFKL.

56. Carter, *US Army before Vietnam*, 38; Trauschweizer, *Cold War U.S. Army*, 162–80.

57. Memorandum 19, from HQ, CONARC, "Staff Organization for the Tactical Employment of Atomic Weapons," July 27, 1957, box 7, entry P 50470 Third Army, G-3 Records Relating to Training 1957–1960, RG 338, NARA; "The U.S. Army Today," n.d. (possibly 1958), folder: Miscellaneous papers of Secretary of the Army, box 17, Brucker Papers; DA Pamphlet 355-13, May 1958; Lewis, "American Culture of War," 290.

58. DA, FM 100-5, *Operations*, 1962, 3-6; Lewis, *American Culture of War*, 219–21.

59. US Senate, *Study of Airpower*, 1618; Gavin, *War and Peace*, 91.

5. Tactical Mobility and the Airmobile Division

1. Robert H. Scales, *Certain Victory: The U.S. Army in the Gulf War* (Washington, DC: Potomac Books, 1997), 218–20; William B. Ostlund, "The Largest Air Assault in History," *The Spear* (podcast), January 29, 2020, prod. John Amble, MP3 audio, 29:45, https://the-spear.castos.com/episodes/the-largest-air-assault-in-history; "Air Assault in the Gulf," oral history interview of Maj. Gen. J. H. Binford Peay III, commanding general, 101st Airborne Division, June 5, 1991, by Robert K. Wright, Rex Boggs, and Cliff Lippard, CMH.

2. James M. Gavin, *War and Peace*, 112–14, 257; Ridgway and Martin, *Soldier*, 311–14; Kinnard Oral History, 147–48.

3. Hanson W. Baldwin, "The Cavalry Charges On," *New York Times*, July 2, 1950.

4. Gavin, "Cavalry, and I Don't Mean Horses," 59.

5. See James M. Gavin to Hamilton Howze, April 13, 1959, folder 3, box 1, Howze-Hawkins Family Papers, USAHEC; Gavin, "Cavalry, and I Don't Mean Horses," 54–55; Gavin, *War and Peace*, 216–18.

6. Gavin, *War and Peace*, 109, 137, 271–72.

7. "Army Building Up Own Airlift Force," *New York Times*, November 16, 1952; address by the Under Secretary of the Army Earl D. Johnson before the Air Power Preparedness Symposium of the Seventh Annual National Convention of the Air Force Association, Washington, DC, August 21, 1953, and address by Johnson before Dallas Council on World Affairs Symposium on Air Power, Dallas, TX, November 20, 1953, box 7, Earl D. Johnson Papers, HSTL.

8. "Airmobility for the New Army," speech by Frank Pace, Secretary of the Army, May 15, 1952, before the American Helicopter Society, Washington Hotel, Washington, DC, entry UD 390-D, box 293, RG 407, NARA; Matthew B. Ridgway, testimony before Subcommittee on Defense Appropriations, February 7–8, 1955, 23.

9. Gavin, *War and Peace*, 44; Gavin Oral History, section 1, pp. 28–29; Gavin, "Cavalry, and I Don't Mean Horses," 60.

10. Col. Joseph W. Stilwell Jr., Comments, Infantry Instructors Conference, June 21–26, 1954, Report, Donovan Research Library, Fort Benning, GA; Linn, *Elvis's Army*, 111; James J. Haggerty, "No More Paratroops?," *Colliers*, March 18, 1955, 23; Carl I. Hutton, "The Commandant's Column: An Air Fighting Army?," *Army Aviation Digest*, July 1955, 2.

11. Howze Oral History, 52; Hamilton H. Howze, "The Army's Stake in the Helicopter," address at the 12th Annual National Forum of the American Helicopter Society, May 4, 1956, 29-31, Howze-Hawkins Family Papers, box 8, folder 2, USAHEC; presentation by Maj. Gen. Hamilton Howze, Director of Army Aviation, ODCSOPS, August 29, 1957, folder 171, box 63, Baldwin Papers; "Air Cavalry," presentation by Director of Army Aviation, December 11, 1957, box 1, Howze Papers; Hamilton H. Howze, *A Cavalryman's Story: Memoirs of a Twentieth-Century Army General* (Washington, DC: Smithsonian Institution, 1996), 233-36.

12. Howze, *Cavalryman's Story*, 184, 230; Frederic A. Bergerson, *The Army Gets an Air Force: Tactics of Insurgent Bureaucratic Politics* (Baltimore: Johns Hopkins University Press, 1980), 51.

13. John J. Tolson, *Airmobility 1961-1971* (Washington, DC: GPO, 1973); John J. Tolson, "The Future of Army Aviation," Army War College student paper (April 1, 1953), 15, 28-30, USAHEC; HQ, Infantry School, Memorandum No. 1, "Reorganization of the Airborne Department," February 11, 1955, Fort Benning, GA, folder 181, box 63, Baldwin Papers; John Norton Oral History Interview, John Norton Collection (AFC/2001/001/31599), Veterans History Project, American Folklife Center, LOC.

14. Stewart L. McKenney, "SKYCAV Operations during Exercise Sagebrush," *Military Review* 36, no. 3 (June 1956): 12-18; Observer Report on SAGEBRUSH, Col. John Tolson, Director of Airborne-Army Aviation Department (December 9, 1955), p. 2, entry 109 (A1), box 5, folder 3, RG 546, NARA; "Army Looks to Sage Brush's SkyCav Operation as Basis of Permanent Unit," *Army-Navy-Air Force Journal and Register*, December 10, 1955, 1, 3; Maj. Robert Slepian, Observer Report, Exercise Sagebrush, December 1, 1955, p. 4, entry 109 (A1), box 5, folder 3, RG 546, NARA; Linn, *Elvis's Army*, 112.

15. Linn, *Echo of Battle*, 174; Paul D. Adams, "Final Report, Exercise Sage Brush," January 1956, 3-5, Reference Material box 5, Training Exercises, WMM82.

16. F. H. Carten to Brig. Gen. Ghormley, April 2, 1972, box 4, Vanderpool Papers; Jay Vanderpool, "We Armed the Helicopter," *Army Aviation Digest* 17, no. 6 (June 1971): 29; Jay D. Vanderpool Oral History, interview by Lt. Col. John R. McQuestion, 1983, p. 158, box 1, Vanderpool Papers; Tolson, *Airmobility*, 5; multiple trip reports from Army Aviation School personnel going to the Infantry School at Fort Benning between 1956 and 1967, see box 10, entry UD 7, Aviation School, Fort Rucker, AL, 1953-63, RG 546, NARA; Lt. Col. Bennett to CG, United States Continental Army Command, memorandum, "Establishment of an Experimental Aerial Battalion (Infantry), June 24 1958, DCSOPS Decimal Files, 1956-1964, 1958 records, entry P 218 Central Decimal Files, RG 319, NARA.

17. Jay D. Vanderpool, "Initial Concepts, Approaches, and Reasons for Arming the Helicopter," paper given at Army Scientific Advisory Panel meeting, May 14, 1969, Fort Rucker, AL, box 4, Vanderpool Papers; Vanderpool, "We Armed the Helicopter," 5-6; Department of the Army, *Army Transport Aviation Combat Operations*, Field Manual 57-35 (Washington, DC: GPO, 1958); Vanderpool Oral History, 166-67; Shelby Stanton, *Anatomy of a Division: 1st Cav in Vietnam* (Novato, CA: Presidio, 198, 13.

18. Joseph O. Wintersteen Jr., "Helicopterborne Operations," *Infantry Journal* 47, no. 2 (April 1957): 22–33; Robert McMahon, "Airmobile Operations," *Military Review* 33, no. 3 (June 1959): 32; Department of the Army, *Airmobile Operations*, Field Manual 57-35 (Washington, DC, 1960).

19. J. D. Coleman, *Pleiku: The Dawn of Helicopter Warfare in Vietnam* (New York: St. Martin's, 1988), 4; Enclosure I to Section VII, "The Requirements for Air Fighting Unit," Army Aircraft Requirements Review Board, reprinted in Howze, *Cavalryman's Story*, 235; Tolson, *Airmobility*, 8; letter to Army Chief of Staff, Subj: Army Aircraft Requirements Review Board, March 10, 1960, in Stanton, *Anatomy of a Division*, 13; Mark D. Sherry, "Airmobility," in *A History of Innovation*, ed. Jon T. Hoffman (Washington, DC: CMH, 2009), 118.

20. Memorandum for Mr. Stahr, April 19, 1962, reprinted in Report of US Army Tactical Mobility Requirements Board, Fort Bragg, NC, August 20, 1962, p. 15, USAHEC; Gregory A. Daddis, *No Sure Victory: Measuring U.S. Army Effectiveness and Progress in the Vietnam War* (New York: Oxford University Press, 2011), 75.

21. Robert R. Williams Oral History Interview, May 8, 1984, Robert R. Williams papers, USAHEC.

22. Expanded Outline—1st Air Cav in Vietnam, p.1, box 5, folder 10, J. D. Coleman Collection, TTU Vietnam Archive (hereafter Coleman Collection); Secretary of the Army to Chief of Staff, message, Subj: "Terms of Ref for Air Force Liaison Officer with the Howze Board," reel 391, image 846, Decimal Sheets, January 1961–December 1962, 334 Boards, Commissions, Committees, Councils, and Missions, Copies of Cross Reference Sheets to General Correspondence, 1947–1964, RG 335, NARA.

23. Maj. Gen. Creighton Abrams to Lt. Gen. Hamilton Howze, June 20, 1962; James M. Gavin to Hamilton Howze, June 18, 1962; William C. Westmoreland to Hamilton Howze, July 11, 1962, all in box 1, George W. Putnam Papers, USAHEC. See same box for further responses to Howze's requests for information.

24. 82nd Airborne Division Training Memorandum No. 10, "Army Training Test for Airborne Division Battle Group," March 22, 1962, folder 9, box 1, Howze Board, Army Tactical Mobility Requirements Field Tests, Edward L. Rowny Papers, USAHEC (hereafter Rowny Papers); General Orders No. 103, June 11, 1962, box 62, WMM82; Eugene M. Zuckert, Secretary of the Air Force, to Secretary of the Army, message, Subj: "Army Tactical Mobility Requirements Board," June 15, 1962, reel 391, image 848, Decimal Sheets, January 1961–December 1962, RG 335, NARA; Howze, *Cavalryman's Story*, 239; Finn J. Larsen, Assistant Secretary of the Army for Research and Development, multiple messages to industry leaders, June 1–12, 1962, reel 391, image 852–55; Decimal Sheets, January 1961–December 1962, RG 335, NARA.

25. Testing Schedule, box 1, Howze Board, Army Tactical Mobility Requirements Field Tests, Rowny Papers; US Army Combat Developments Command, *The Origins, Deliberations, and Recommendations of the U.S. Army Tactical Mobility Requirements Board* (Fort Leavenworth, KS, 1969), 50–51; Howze, *Cavalryman's Story*, 243–50; Correspondence from Cyrus Vance regarding testing, July 6, 1962, folder 6, box 1A, Howze-Hawkins Family Papers, USAHEC.

26. Report of US Army Tactical Mobility Requirements Board, Fort Bragg, NC, August 20, 1962, p. 95, USAHEC.

27. Expanded Outline–1st Air Cav in Vietnam, p. 2, box 5, folder 10, Coleman Collection; Summary Sheet, Test Plan 2-2, "Basic Air Mobility Force," folder 8, box 1, Rowny Papers, USAHEC.

28. Report of US Army Tactical Mobility Requirements Board, Fort Bragg, NC, August 20, 1962, USAHEC; Expanded Outline–1st Air Cav in Vietnam, p. 2, box 5, folder 10, Coleman Collection; Tolson, *Airmobility*, 22.

29. Report of US Army Tactical Mobility Requirements Board, Fort Bragg, NC, August 20, 1962, USAHEC; Stanton, *Anatomy of a Division*, 20; Howze, *Cavalryman's Story*, 253.

30. Defense budget FY 64, vol. 1: Recommended General Purpose Forces, December 4, 1962, pp. 29–32, Presidential Papers, National Security Files, Department of Defense, Papers of John F. Kennedy, JFKL.

31. Defense budget FY 64, vol. 1: Recommended General Purpose Forces, December 4, 1962, p. 32, JFKL; Chief of Staff of the Army from Secretary Vance, memorandum, Subj: "Appreciation," February 8, 1963, folder 6, box 1A, Howze-Hawkins Papers; Howze, *Cavalryman's Story*, 254–56; Vanderpool Oral History, 181.

32. Stanton, *Anatomy of Division*, 24–27; Hamilton Howze, "Tactical Employment of the Air Assault Division," *Army* 14, no. 2 (September 1963): 35; Christopher C. S. Cheng, *Air Mobility: The Development of a Doctrine* (Westport, CT: Praeger, 1994), 82.

33. Department of the Army, *Army Mobility Concept* (Washington, DC: GPO, December 1963), II-23, III-D-5; Harold Moore and Joseph L. Galloway, *We Were Soldiers Once . . . and Young: Ia Drang—the Battle That Changed the War in Vietnam* (New York: Ballantine Books, 2004), 11; Stanton, *Anatomy of a Division*, 27.

34. Kinnard Oral History, 44–49, 62–65; Wellman, *Battleground!*; Lt. Col. George F. Charlton, memorandum to DCOS-Training and Readiness, CONARC, "Report of Staff Visit to 2d Infantry Division, 21 May 1964," May 27, 1964, box 23, entry UD WW-125, General Records, CONARC HQ 1957-66 Unclassified, RG 546, NARA; Moore and Galloway, *We Were Soldiers Once*, 11; Stanton, *Anatomy of a Division*, 22–24; J. D. Coleman, *Pleiku: The Dawn of Helicopter Warfare in Vietnam* (New York: St. Martin's, 1988), 12.

35. Kinnard quoted in Coleman, *Pleiku*, 22.

36. Coleman, *Pleiku*, 21; Moore and Galloway, *We Were Soldiers Once*, 17.

37. Coleman, *Pleiku*, 22; Moore and Galloway, *We Were Soldiers Once*, 20.

38. Coleman, *Pleiku*, 21; Stanton, *Anatomy of a Division*, 34; "Qualification Badges, Air Assault Badge," U.S. Army Institute of Heraldry, dated January 18, 1978, https://tioh.army.mil/Catalog/Heraldry.aspx?HeraldryId=15435&Categ oryId=9361&grp=2&menu=Uniformed%20Services&ps=24&p=0.

39. Stanton, *Anatomy of a Division*, 29–30; Tolson, *Airmobility*, 52; Cyrus R. Vance, Secretary of the Army, address before the Army Tactical Mobility Symposium, November 19, 1963, Fort Benning, GA, folder 49, box 5, Cyrus R. and Grace Sloane Vance Papers (MS 1664), Manuscripts and Archives, Yale University Library (hereafter Vance Papers).

40. Gavin Oral History, 29–30; Stanton, *Anatomy of a Division*, 32–33.

41. Quoted in Tolson, *Airmobility*, 58; Itinerary for Chief of Staff Visit, November 4–5, 1964, Fort Jackson, SC, folder 20, box 138, Harold K. Johnson papers, USAHEC.

42. Harry W. O. Kinnard, "Airmobility Revisited, Part 2," *Army Aviation Digest* 26, no. 7 (July 1980): 8; Daddis, *No Sure Victory*, 76; Gen. Earle G. Wheeler, testimony to Congress, in "The Prospects for Army Air Mobility," *Army*, March 1963, 20; Cheng, *Air Mobility*, 185–87; Brig. Gen. John Tolson to Hanson Baldwin, December 3, 1964, folder 860, box 17, Baldwin Papers; Gen. Earle G. Wheeler, "Army Moves toward Mobility," *Army Information Digest*, February 1964, 34–35.

43. Tolson, *Airmobility*, 52–54; Moore and Galloway, *We Were Soldiers Once*, 15–16.

44. Hanson Baldwin, "Strategy in Vietnam," *New York Times*, September 21, 1963; Bernard B. Fall, *Street without Joy: The French Debacle in Indochina* (Guilford, CT: Stackpole Books, 2018), 360–61.

45. Kinnard Oral History, 148; Office of the Assistant Secretary of Defense (Public Affairs), press release no. 404–54, June 16, 1965, folder 181, box 63, Baldwin Papers; Richard Goldstein, "Gen. H. H. Howze, 89, Dies; Proposed Copters as Cavalry," *New York Times*, December 18, 1998.

46. Lewis B. Sorley, *Honorable Warrior: General Harold K. Johnson and the Ethics of Command* (Lawrence: University Press of Kansas, 1998), 91–111.

47. Memorandum of Agreement between Gen. Harold K. Johnson, chief of staff, US Army, and Gen. John P. McConnell, chief of staff, US Air Force, April 6, 1966, in Roy L. Bowers, *Tactical Airlift*, The United States Air Force in Southeast Asia (Washington, DC: Office of Air Force History, 1983), appendix 6, 673–74.

48. Kinnard Oral History, 14; Maj. Gen. Harry W. O. Kinnard, "Activation to Combat—in 90 Days," *Army Information Digest* 21, no. 4 (April 1966): 27; Stanton, *Anatomy of a Division*, 36; Coleman, *Pleiku*, 39.

49. Robert S. McNamara, secretary of defense, report, "Deployment of Major US Forces to Vietnam, July 1965, Vol. 7," July 23, 1965, National Security Files, LBJL; Lyndon B. Johnson, "Why We Are in Viet-Nam," July 28, 1965, President's News Conference, American Presidency Project, University of California, Santa Barbara, https://www.presidency.ucsb.edu/documents/the-presidents-news-conference-1038; Abrams quoted in Tolson, *Airmobility*, 61; Rusk to Taylor, cable, July 28, 1965, folder, "Deployment of Major U.S. Forces to Vietnam, July 1965, Volume 7, Tabs 421–438," National Security Council Histories, NSF, box 43, LBJL; Kinnard, "Activation to Combat," 30–31.

50. Lt. Gen. Harry W. O. Kinnard, interview, September 19, 1990, folder 6, box 48, Harold G. Moore Papers, USAHEC (hereafter Moore Papers); Tolson, *Airmobility*, 254; Stephen Peter Rosen, *Winning the Next War: Innovation and the Modern Military* (Ithaca, NY: Cornell University Press, 1994), 93; Stanton, *Anatomy of a Division*, 34–35.

51. Maj. Gen. Harry W. O. Kinnard, First Cavalry Division (Airmobile), "Combat Operations After Action Report: Plei Mei Campaign, October 23–November 26, 1965," March 4, 1966, CMH, p. 28; Rosen, *Winning the Next War*, 94; Daddis, *No Sure Victory*, 79.

52. 3rd Brigade, 1st Air Cav Div, FRAGO 65-12 to OPORD 7-65, 13 Nov 65, folder 6A, box 46, Moore Papers; Gregory A. Daddis, *Westmoreland's War: Reassessing American Strategy in Vietnam* (New York: Oxford University Press, 2015), 97; After Action Report, 1st Battalion, 7th Cavalry—IA DRANG Valley Operation, November 14-16, 1965, December 9, 1965, box 01, folder 01, Operation Masher / Operation White Wing Collection, TTU Vietnam Archive, 1st Cavalry Division Tactical Operations Center Duty Log, November 14, 1965, folder 16, box 48, Moore Papers (hereafter 1st Cav Div TOC Duty Log); see also John M. Carland, *Combat Operations: Stemming the Tide, May 1965 to October 1966* (Washington, DC: CMH, 2000), 113.

53. 1st Cav Div TOC Duty Log, November 16, 1965; Capt. Henry B. Thorpe, After Action Report—Operation Silver Bayonet, November 12-20, 1965, Company D, 2-7 Cavalry, 1st Cavalry Division (Airmobile), folder 6A, box 46, Moore Papers.

54. Report, 1st Air Cavalry Division—Lessons Learned from Operations, December 29, 1965, box 4, folder 3, Coleman Collection; Harry W. O. Kinnard, "A Victory in the Ia Drang: The Triumph of a Concept," *Army* 17, no. 9 (September 1967): 89. For a description of the Pleiku campaign and the battles at X-Ray and Albany see Carland, *Combat Operations*, 113-50 (casualty statistics on 145), Tolson, *Airmobility*, 73-83, and Moore and Galloway, *We Were Soldiers Once*; Interview with Lt. Gen. Harry W. O. Kinnard, September 19, 1990, folder 6, box 48, Moore Papers.

55. Kinnard quote in *Newsweek*, December 13, 1965, 28; CIA Weekly Report, "The Situation in South Vietnam," November 24, 1965, folder 6A, box 46, Moore Papers.

56. William C. Westmoreland to Harold Moore, May 15, 1983, folder 6, box 48, Moore Papers; Westmoreland quoted in "Westmoreland Surveys Action," *New York Times*, November 20, 1965; Daddis, *No Sure Victory*, 81; see also Daddis, *Westmoreland's War*, 98; Carland, *Combat Operations*, 150; Tolson, *Airmobility*, 83; Kretchik, *U.S. Army Doctrine*, 188; adviser quoted in Malcom W. Browne, *The New Face of War*, rev. ed. (Indianapolis: Bobbs-Merrill, 1968), 79.

57. Daddis, *No Sure Victory*, 82-85; Johnson in William Conrad Gibbons, *The U.S. Government and the Vietnam War: Executive and Legislative Roles and Relationships, Part IV: July 1965-January 1968* (Princeton, NJ: Princeton University Press, 1995), 101-2; Rosen, *Winning the Next War*, 94; Daddis, *Westmoreland's War*, 98-99; Hayley Michael Hasik, "The Helicopter War: Unraveling the Myth and Memory of a Vietnam War Icon" (PhD diss., University of Southern Mississippi, 2023).

58. David Lamb, "Copter Proves Itself as Vietnam Weapon," *Nashville Tennessean*, July 6, 1969; William C. Westmoreland, *A Soldier Reports* (New York: Doubleday, 1976), 348; Matthew Allen, *Military Helicopter Doctrines of the Major Powers 1945-1992: Making Decisions about Air-Land Warfare* (Westport, CT: Greenwood, 1993), 11; Krepinevich, *Army and Vietnam*, 122.

59. Oral history transcript, Stanley R. Resor, interview 1 (I), November 16, 1968, by Dorothy Pierce, p. 8, LBJ Library Oral Histories, LBJL; Gen. Melvin Zais to Maj. Gen. Sidney Berry, February 25, 1974, Zais to Berry, February 25,

1974, and Berry to Zais, February 4, 1974, all in box 3, Melvin Zais Papers, USA-HEC. Zais and Berry communicated about the loss of parachute slots and the meaning of "airborne." Zais was careful to note that "airborne" was a generic term, denoting glider, parachute, air-land, and helicopter methods of arrival.

60. Wilson, *Maneuver and Firepower*, 359, 368, table 35.

6. The Strategic Army Corps and the Emergence of Strike Command

1. Gavin, *Airborne Warfare*, 170.

2. James M. Gavin, "The Future of Airborne Operations," *Military Review* 27, no. 9 (December 1947): 4–5, 8 (Devers quote on 8); Matthew B. Ridgway, "An Army on Its Toes," *Army Combat Forces Journal*, December 1954, 10; Bacevich, *Pentomic Era*, 106; Linn, *Elvis's Army*, 214.

3. Robert K. Wright, "Airborne Forces and the American Way of War," *Army History* 72 (Summer 2009): 47.

4. Linn, *Elvis's Army*, 304; Wright, "Airborne Forces," 41–42.

5. T. F. Walkowicz, *Future Airborne Armies: A Report Prepared for the Army Air Forces Scientific Advisory Group* (Wright Field, OH: Air Material Command, September 1945), 1–2, 60–63; Lt. Gen. L. H. Brereton, HQ 3rd Air Force, memorandum to HQ Army Air Forces, August 7, 1945, Airborne Operations folder, box 8, Floyd Parks Papers, DDEL; for more on the First Allied Airborne Army see James A. Huston, *Out of the Blue: U.S. Army Airborne Operations in World War II* (West Lafayette, IN: Purdue University Press, 1972), 77–79; Gen. Omar N. Bradley, address to the graduates of the Command and General Staff College, July 1, 1949, Fort Leavenworth, KS, in W. A. Kuhn, "How Far Along Are We in Developing an Airborne Army?," *Military Review*, April 1950, 41; "Approximate 'Air Lift' Required to Transport One Division," n.d. (circa 1949), folder 180, box 63, Baldwin Papers; "First Rate Fighting Unit," from Donald T. Kellett and William Friedman, "Airborne on Paper Wings," *Infantry Journal* 62, no. 5 (May 1948): 9.

6. John B. Spore, "An Army with Wings," *Reporter*, January 8, 1952, 31–34; Maxwell D. Taylor interview with Richard A. Manion, section 2, p. 29, November 10, 1972, Senior Officers Debriefing Program, Maxwell Taylor Papers, USAHEC (hereafter Taylor Oral History; CG, AGF to Army Chief of Staff, November 2, 1945, box 37, Gavin Papers; Gavin to Gen. E. C. Meyer, Chief of Staff of the Army, July 1, 1981, box 27, Gavin Papers; Joseph Rockis, "Reorganization of Army Ground Forces during the Demobilization Period," AGF Demobilization Study #3 (Fort Monroe, VA, 1948), 47; Jack Draper Oral History (AFC/2001/001/09556), Veterans History Project, LOC; David Halberstam, *The Fifties* (New York: Random House, 1993), 70–71; Kellett and Friedman, "Airborne on Paper Wings," 13.

7. Wilson, *Maneuver and Firepower*, 226; Ernest F. Fisher, "Evolution of U.S. Airborne Doctrine," *Military Review*, May 1966, 72–74.

8. Omar N. Bradley, "Creating a Sound Military Force," *Military Review* 24, no. 2 (May 1949): 3–6; Ridgway, "Trends in Modern Warfare," 6; Melvin Zais to William Miley, "Employment of Airborne Troops," February 1, 1950, Airborne Historical Studies, box 1, Zais Papers; "Organization, Equipment, and Tactical

Employment of the Airborne Division," Study No. 16, Reports of the General Board, US Forces, European Theater, CARL.

9. Report to the President, "NATO Force Goals," April 25, 1952, box 13, National Security File, Harry S. Truman Papers, HSTL; Ridgway letter, April 22, 1954, box 4, NATO Series, Alfred M. Gruenther Papers, DDEL.

10. Daniel F. Harrington, *Berlin on the Brink: The Blockade, the Airlift, and the Early Cold War* (Lexington: University Press of Kentucky, 2012), 232; Kuhn, "How Far Along Are We?," 44; Owen, *Air Mobility*, 84–85; Jacob L. Devers, "Air Transportability of the Infantry Division," *Military Review* 24, no. 1 (April 1949): 14–18; Lt. Col. Zais to Col. Berquist, "Employment of Airborne Troops," January 24, 1950, Airborne Historical Studies, box 1, Zais Papers.

11. Austin Stevens, "Planes Fall Short of Airlift Needs," *New York Times*, May 2, 1950; Ridgway to Vice Chief of Staff, "Movement of 2-Division Corps to Europe," May 26, 1950, box 76, Ridgway Papers; Mathewson to Ridgway, August 29, 1950, box 31, Ridgway Papers; Ridgway and Martin, *Soldier*, 312; Kellett and Friedman, "Airborne on Paper Wings," 9.

12. "Final Report, Exercise Swarmer," Training Operations Files for Maneuver Exercise "Swarmer," 1/1950–5/1950, box 800, entry UD-UP 1, RG 337, NARA; 11th Airborne Division Exercise Swarmer Final Report, 1950, entry UD 37042, Unit Histories, Division Section, 1943–67, RG 338, NARA; Albert Pierson, "Airborne Operations," *Army Information Digest* 9, no. 7 (July 1954): 20–30; Command Report, Exercise Swarmer, 82nd Airborne Division, box 4388, entry NM3 429, US Army Command Reports, 1948–1954, 429, RG 407, Records of the Adjutant General, NARA; Final Report, Exercise Swarmer, 1950, 82nd Airborne Division, entry UD 37042, Unit Histories, Division Section, 1940–67, RG 338, NARA; Hunter M. Brumfield, "Swarmer—a Pattern for Airborne Assault," *Army Information Digest* 5, no. 7 (July 1950): 13–22; see also "Airpower Becomes Supply Power," *Life*, May 15, 1950, 42–43.

13. After Action Report, CG, 187th RCT, October 29, 1950, 11th Airborne Division, box 3413, entry NM3 429, US Army Command Reports, 1948–1954, 429, RG 407, NARA; Owen, *Air Mobility*, 95–96; J. Lawton Collins to US House of Representatives Subcommittee on Department of the Army Appropriations, 82nd Congress, 2nd session, February 5, 1952.

14. Bolte to Ridgway, January 19, 1950, box 6, Ridgway Papers; CS 2045/8, April 26, 1951, Records of the Air Force Representative on the Joint Airborne Troop Board, RG 340, NARA; Galvin, *Air Assault*, 264.

15. See Joint Airborne Troop Board Administrative and Project Files, 1950–1954, RG 337, NARA; Report of the Army Airborne Conference, 1951, 1953 Airborne Conference Report, Joint Airborne Troop Board, Fort Bragg, and Report of the 1957 Army Airborne Conference, Fort Bragg, all in USAHEC. See also Spore, "Army with Wings," 31–34; 1953 Airborne Conference Report, p. 5, November 17–20, 1953, Joint Airborne Troop Board, Fort Bragg, USAHEC. For more on the development of joint doctrine see Robert Frank Futrell, *Ideas, Concepts, Doctrine*, vol. 1, *Basic Thinking in the United States Air Force, 1907–1960* (Maxwell Air Force Base, AL: Air University Press, 2004), 378–79.

16. Address by the Under Secretary of the Army Earl D. Johnson before the Air Power Preparedness Symposium of the 7th Annual National Convention of

the Air Force Association, Washington, DC, August 21, 1953, and Address by Under Secretary of the Army Earl D. Johnson before Dallas Council on World Affairs Symposium on Air Power, Dallas, TX, November 20, 1953, box 7, Earl D. Johnson Papers, HSTL.

17. Appendix A1-2, Report of the Army Airborne Conference, February 19-23, 1951, Army Airborne Center, Fort Bragg, USAHEC; Taylor Oral History, section 2, pp. 29-30; Maxwell D. Taylor, "The Changing Army," *Army Forces Combat Journal*, October 1955, 10.

18. Secretary Wilbur M. Brucker, Address at Lincoln Dinner, Carlisle, PA, February 11, 1956, box 27, Brucker Papers; Ridgway to Secretary of Army, June 27, 1955, Official Correspondence (Army Chief of Staff), box 18, Ridgway Papers.

19. Earle Wheeler, "Strategic Mobility," *Army Information Digest* 12, no. 1 (January 1957): 2-12; 82nd Airborne Division, "Report of Students Jumping from C-124 Aircraft," memorandum, May 29, 1957, box 7, entry P 50470 Third Army, G-3 Records Relating to Training 1957-1960, RG 338, NARA; Department of Defense, *Semiannual Report of the Secretaries of the Army, Navy, and Air Force, January 1 to June 30, 1956* (Washington, DC: GPO, 1956), 268; E. R. Johnson and Lloyd S. Jones, *American Military Transport Aircraft since 1925* (Jefferson, NC: McFarland, 2013), 209-17.

20. Maxwell D. Taylor, Address at Tenth Annual Convention, National Defense Transportation Association, Sheraton Plaza Hotel, Boston, "The Army and Mobility," October 13, 1955, Taylor Papers, NDU; Charles E. Wilson, speech to Republican Party of Massachusetts, June 5, 1957, box 35, Wilson Papers.

21. "Renaming the Strategic Reserve," March 8, 1957, box 234, entry P 218, Records of the DCoS for Military Operations, Central Decimal Files, 1956-1964, 1957 records, RG 319, NARA; Department of Defense, *Semiannual Report of the Secretary of the Army*, October 13, 1958, p. 103, box 784, Office of the Secretary of the Army, General Correspondence January 1957-December 1960, RG 335, NARA; James Gavin, "We Can Solve Our Technical Difficulties," *Army Combat Forces Journal*, November 1955, 64; "Organizing and Deploying Our Forces," *Army Information Digest* 13, no. 6 (June 1958): 28-29; DCSOPS to DC-SLOG, memorandum, "Strategic Army Forces," February 19, 1957, and memorandum for DCSOPA, "STRAF and STRAC Divisions under August 17, 1957 Structure," box 234, entry P 218, Records of the DCoS for Military Operations, Central Decimal Files, 1956-1964, 1957 records, RG 319, NARA.

22. "Improved Readiness of the STRAC," memorandum, June 20, 1957, entry P 218, box 234, Records of the DCoS for Military Operations, Central Decimal Files, 1956-1964, 1957 records, RG 319, NARA; "Military Objectives of the Army, 1960 to 1962," address by Maxwell D. Taylor at the Secretaries' Conference, Quantico, VA, June 21, 1958, Taylor Papers, NDU; "STRAC–Mobile Striking Force in Readiness," *Army Information Digest* 13, no. 6 (August 1958): 16-17; Galvin, *Fighting the Cold War*, 102-3.

23. Linn, *Elvis's Army*, 225-26; Howze Oral History, section 5, p. 5, Howze Papers; Robert Haldane Oral History, 38-39, box 1, Robert Haldane Papers, USAHEC; Third Army G3 Training Division, memorandum, "Readiness of

STRAC Forces," July 21, 1960, and memorandum, "Deployment Readiness of STRAC Units," July 27, 1960, box 23, entry P 50470 Third Army, G-3 Records Relating to Training 1957–1960, RG 338, NARA; Hanson W. Baldwin, "The New Army Corps," *New York Times*, May 24, 1958.

24. Norman H. Schwarzkopf with Peter Petre, *It Doesn't Take a Hero* (New York: Bantam Books, 1992), 73–84, quote on 73; DCSOPS to DCSLOG, memorandum, "Strategic Army Forces," February 19, 1957, box 233, entry P 218, Records of DCoS for Military Operations, Central Decimal Files, 1956–1964, 1957 records, RG 319, NARA.

25. Telephone calls log, September 25, 1957, box 27, DDE Diary Series, Dwight D. Eisenhower Papers as President (Ann Whitman file), DDEL; George M. Seignious, memorandum for record, Subj: "Use of Federal Troops in Little Rock, Arkansas," October 28, 1957, box 17, Brucker Papers, BHL; information received from Gen. Wheeler, 24 2045 Sep 1957, box 1, Command Report, Operation ARKANSAS, box 2, entry UD 116, Records of the Office of the DCoS for Military Operations, Domestic Disturbance Files, Records of Operation Arkansas, 1957–1958, RG 319, NARA; Paul J. Scheips, *The Role of Federal Military Forces in Domestic Disorders, 1945–1992* (Washington, DC: CMH, 2005), 40, 44.

26. Brig. Gen. Goodpaster, memorandum for record, May 15, 1958, Department of Defense vol. 2, May 1958, box 1, White House Office of the Staff Secretary, Subject Series: DoD Subseries, DDEL; Tad Szulc, "US Flies Troops to Caribbean as Mobs Attack Nixon in Caracas," *New York Times*, May 14, 1958; William I. Hitchcock, *The Age of Eisenhower: America and the World in the 1950s* (New York: Simon & Schuster, 2019), 418.

27. Department of Defense, *Semiannual Report of the Secretary of the Army* (Washington, DC: GPO, 1958), 104; dispatch from DA, Washington, DC, to CG, USCONARC, CCS 381, Combined Plans for the Defense of the Western Hemisphere, Sec. 38, Geographic Files, 1958, RG 218, NARA; Robert Sink, "STRAC Plans for the Future," *Army Information Digest* 14, no. 11 (November 1959): 2; dispatch from DA, Washington, DC, to CG, 101st Abn Div Task Force, CCS 381, Combined Plans for the Defense of the Western Hemisphere, Sec. 38, Geographic Files, 1958, RG 218, NARA.

28. Memorandum for Gen. Goodpaster, May 17, 1958, Department of Defense vol. 2, May 1958, box 1, White House Office of the Staff Secretary, Subject Series: DoD Subseries, DDEL; Department of the Army, "1957 Airborne Conference," appendix B3, p. 252, USAHEC.

29. Chairman's Staff Group, memorandum, "Service Airlift Emergency Plans," May 20, 1958, Department of Defense vol. 2, May 1958, box 1, White House Office of the Staff Secretary, Subject Series: DoD Subseries, DDEL.

30. Jonathan M. House, *A Military History of the Cold War, 1944–1962* (Norman: University of Oklahoma Press, 2012), 353–54; Gary H. Wade, *Rapid Deployment Logistics: Lebanon, 1958* (Fort Leavenworth, KS: CSI Press, 1984), 8–9; Lebanon Situation, CCS 381 Lebanon Sec. 1, Geographic Files, 1958, box 10, RG 218, NARA; Randall Fowler, *More Than a Doctrine: The Eisenhower Era in the Middle East* (Lincoln, NE: Potomac Books, 2018), 139–42.

31. Department of Defense, *Semiannual Report of the Secretary of the Army* (Washington, DC: GPO, 1958), 104–5; Roger J. Spiller, *"Not War but Like War"*:

The American Intervention in Lebanon (Fort Leavenworth, KS: CSI Press, 1981), 9–10, 26; dispatch from CINCSPECOMME to CNO, "For Admiral Burke. Sitrep 12. Blue Bat," Lebanon Situation, CCS 381 Lebanon Sec. 4, Geographic Files, 1958, box 10, RG 218, NARA; Neil McElroy oral history interview with Philip A. Crowl, May 6, 1964, John Foster Dulles Oral History Project, MC017, Public Policy Papers, Mudd Manuscript Library, Princeton University; Gordon Rottman, *US Army Airborne, 1940–1990* (New York: Osprey, 1990), 29; Birtle, *US Army Counterinsurgency*, 190.

32. Gen. Lyman L. Lemnitzer, "The Army & National Security," September 24, 1959, at the Sixth Annual Conference of Civilian Aides to the Secretary of the Army, September 23–25, 1959, folder: Papers, 1959, box 17, Brucker Papers, BHL; Lt. Gen. Robert F. Sink, "The Strategic Army Corps," September 24, 1959, at the Sixth Annual Conference of Civilian Aides to the Secretary of the Army, September 23–25, 1959, folder: Papers, 1959, box 17, Brucker Papers, BHL.

33. See Final Report, Exercise Eagle Wing, May 13, 1958, box 27, Exercise White Cloud, December 8, 1958, box 12, and Final Report, Exercise Dark Cloud / Pine Cone II, February 16, 1959, box 39, all in entry P 50470, Third Army, G-3 Records Relating to Training 1957–1960, RG 338, NARA. Final Report, Exercise Test Drop, Exercise Snow Storm, Camp Drum, NY, January–March 1953; General Plan, Exercise Flash Burn, April–May 1954; and Final Report, Exercise Arctic Knight, May 11, 1956, all in box 57, entry UD 35176: General Records [XVIII (Airborne) Corps, 1951–1963], RG 338; Jack Raymond, "Paratroop Tests in 'Combat' Hailed," *New York Times*, February 23, 1959; Hamilton H. Howze, "STRAC Flexes Its Muscles," *Army Information Digest* 14, no. 7 (July 1959): 14–23.

34. Report, Subj: "Strategic Mobility Exercises," February 4, 1960, box 1, entry A1 55-E, Records of the Office of the Chief of Information, Security Correspondence, 1960, 62–64, RG 319, NARA.

35. Sink, "STRAC Plans for the Future," 2–10.

36. Lt. Gen. T. J. H. Trapnell to Gen. Herbert B. Powell, January 12, 1961, box 1, Paul D. Adams Papers, USAHEC; for more on Trapnell's career see his obituary: "Thomas Trapnell, 99; Bataan Hero, Military Advisor," *Los Angeles Times*, February 16, 2002.

37. Douglas C. Lovelace and Thomas-Durrell Young, "Defining US Atlantic Command's Role in the Power Projection Strategy," US Army Strategic Studies Institute, August 1998, 3; Ronald H. Cole et al., "The History of the Unified Command Plan 1946–1993," Washington, DC: Office of the Chairman of the Joint Chiefs of Staff Joint History Office, February 1995, 32–35; Organization and Planning Files, USCONARC/ARSTIKE Ops, box 8, entry UD WW-125, General Records, CONARC HQ 1957–66 Unclassified, RG 546, NARA; Scheips, *Role of Federal Military Forces*, 8; Departments of Defense report, "Appraisal of Capabilities of Conventional Forces," May 12, 1961, Papers of John F. Kennedy, Presidential Papers, President's Office Files, JFKL; Rottman, *US Army Airborne*, 36.

38. Galvin, *Fighting the Cold War*, 86, 90; Judson J. Conner, "DRF—Dynamic Deterrent," *Army Information Digest*, March 1965, 38–41; Raymond Weaver Oral History (AFC/2001/001/20967), Veterans History Project, American Folklife Center, LOC.

39. House, *Military History of the Cold War*, 419–21; Howze, *Cavalryman's Story*, 261; Michael Dobbs, *One Minute to Midnight: Kennedy, Khrushchev, and Castro on the Brink of Nuclear War* (New York: Vintage Books, 2008), 104–5, 176.

40. Department of State Circular 549, September 21, 1963; Telegram 193, from Munich to Secretary of State, October 31, 1963, both in Papers of John F. Kennedy, Presidential Papers, National Security Files, Department of Defense (B): Subjects: Military exercises, Big Lift, JFKNSF-279-001, JFKL.

41. Bruce Palmer Jr., *Intervention in the Caribbean: The Dominican Crisis of 1965* (Lexington: University Press of Kentucky, 1989), 3, 30–35; Dominican Republic, Operation "Power Pack," n.d., Research Material, ROAD, box 5, WMM82; Lawrence M. Greenberg, *United States Army Unilateral and Coalition Operations in the 1965 Dominican Republic Intervention* (Washington, DC: CMH, 1987), 37.

42. Drew Brooks, "Obscure Fight in Dominican Republic Taught 82nd Airborne Urban Warfare Lessons," *Fayetteville (NC) Observer*, April 25, 2015; "Authority of the United States to Take Action in the Dominican Republic," memorandum, folder, "McGeorge Bundy, Vol. 10, April 15–May 31, 1965 [2 of 2]," Memos to the President, NSF, box 3, LBJL.

43. Raymond Weaver Oral History, Raymond Weaver Collection (AFC/2001/001/20967), Veterans History Project, American Folklife Center, LOC; Brooks, "Obscure Fight"; Nadia Schadlow, *War and the Art of Governance: Consolidating Combat Success into Political Victory* (Washington, DC: Georgetown University Press, 2017), 191–92, 194; Joint Information Office, HQ US Forces Dominican Republic, Crisis in the Dominican Republic, n.d., Dominican Republic box 1, WMM82; Brooks, "Obscure Fight"; Palmer, *Intervention in the Caribbean*, 137; 82nd Airborne Division, Summary of Activities, 1965, Research Materials, Division AHS, box 54, WMM82.

44. Weaver Oral History, LOC; Hanson W. Baldwin, "Airborne Troops Filling Ranks Depleted by War," *New York Times*, August 2, 1966; Airborne Alert Project, Commanding General's Guidance, ROAD Reference Materials box 5, WMM82; Rottman, *US Army Airborne*, 33–34, 49.

45. 3rd Brigade Annual History, Vietnam, Vietnam box 1, WMM82; Wright, "Airborne Forces," 44. For a detailed account of the Golden Brigade in Vietnam see Robert J. Dvorchak, *Golden Brigade: The Untold Story of the 82nd Airborne in Vietnam and Beyond* (Indianapolis: IBJ, 2020).

46. Final Rpt, Seventh Army Operations on Coal Emergency, 10 Dec 46 (P&O, 004.07, Case 13), RG 319, NARA; Beaumont, "Airborne: Life Cycle of a Military Subculture," 55.

47. "Some suggested questions for Mr. Vance relating to General Abrams' role in civil disturbances," n.d., folder 81, box 20, Vance Papers; Scheips, *Role of Federal Military Forces*, 120; House, *Military History of the Cold War*, 427–28.

48. AAR, Task Force Detroit, August 2, 1967, Reference Materials, ROAD, box 5, WMM82; Scheips, *Role of Federal Military Forces*, 185–99.

49. See books 1 through 3, entry A1-1595, DA Office of the Chief of Staff, Records Relating to Civil Disturbances, 1968, RG 319, NARA; Instructions to Task Force Commanders, book 1, box 1, entry A1-1595, DA Office of the Chief of Staff, Records Relating to Civil Disturbances, 1968, RG 319, NARA;

Information Office release, April 13, 1968, Vietnam box 14, and AAR, TF Washington, April 1968, Reference Materials, ROAD, box 5, WMM82.

50. Message, JCS 7848 to USCINCEUR, 11 Aug 64, Reference Materials, ROAD, box 5, WMM82; 82nd Airborne Division History, 1917–Present (1995), Reference Materials, Division AHS, box 56, WMM82.

51. Howze Oral History, section 2, p. 2; Gavin, *Airborne Warfare*, 175.

52. Linn, *Elvis's Army*, 213, 225.

Epilogue

1. Christopher Donahue, "We Do Not 'Heavy Breathe' In the 82nd Airborne Division," *18th Airborne Corps Podcast*, episode 37, June 7, 2021, prod. Joe Buccino, https://thedoomsdayclock.podbean.com/e/episode-37-we-do-not-heavy-breathe-in-this-division-a-discussion-with-maj-gen-chris-donahue/.

2. Roger Beaumont, "Airborne: Life Cycle of a Military Subculture," *Military Review* 51, no. 6 (June 1971): 53–61; Cockerham, "Selective Socialization," 216; Linn, *Elvis's Army*, 304.

3. Taylor to Rusk, for the president, January 6, 1965, section 5 of 5, ProQuest History Vault (PQV), http://congressional.proquest.com/histvault?q=003221-002-0001; "Report of the Office of the Secretary of Defense Vietnam Task Force," United States–Vietnam Relations, 1945–1967 (the Pentagon Papers), Part IV.B.1, Evolution of the War, the Kennedy Program and Commitments: 1961, P 760, RG 330, NARA; Trauschweizer, *Maxwell Taylor's Cold War*, 152–55; Shelby Stanton, *Vietnam Order of Battle* (Mechanicsburg, PA: Stackpole Books, 2003).

4. Rosen, *Winning the Next War*, 94; Boyne, *How the Helicopter Changed Modern Warfare*, 132.

5. DA, FM 100–5, *Operations*, 1976; DA, FM 100–5, *Operations*, 1982.

6. "A Flotilla of Army Helicopters Joins Attack on Karbala," *Washington Post*, March 29, 2003, https://www.washingtonpost.com/archive/politics/2003/03/29/a-flotilla-of-army-helicopters-joins-attack-on-karbala/3b318b62-7251-406f-b762-2cf0e8822df1/.

7. Michael R. Gordon, "Heading Back to Iraq for Round 2," *New York Times*, March 1, 2004, https://www.nytimes.com/2004/03/01/international/worldspecial3/heading-back-to-iraq-for-round-2.html; Joseph K. Maddry et al., "Impact of Prehospital Medical Evacuation (MEDEVAC) Transport Time on Combat Mortality in Patients with Non-compressible Torso Injury and Traumatic Amputations: A Retrospective Study," *Military Medical Research* 5, no. 1 (June 30, 2018): 22, https://doi.org/10.1186/s40779-018-0169-2; John Ryan and Tony Dokoupil, "In Afghanistan's 'Valley of Death' a Medevac Team's Miracle Rescue," *Newsweek*, November 5, 2012, https://www.newsweek.com/afghanistans-valley-death-medevac-teams-miracle-rescue-63779.

8. FM 100–5 was renumbered FM 3–0 in 2001. For more on the US Army's counterinsurgency focus before and during Vietnam see Daddis, *Westmoreland's War*; DA, FM 100–5, *Operations*, 1976, 2–10.

9. See James Gavin, letter to Michael Cannon, July 20, 1983, box 31, Gavin Papers; DA, FM 100–5, *Operations*, 1962, and DA, FM 100–5, *Operations*, 1976.

For more on this see David Fitzgerald, *Learning to Forget: US Army Counterinsurgency Doctrine and Practice from Vietnam to Iraq* (Stanford, CA: Stanford Security Studies, 2014); John Norton Oral History Interview, John Norton Collection (AFC/2001/001/31599), Veterans History Project, American Folklife Center, LOC.

10. Carter, "Eisenhower versus the Generals," 1196–99; Department of Defense Reorganization Act of 1958, August 6, 1958, in Cole, *Department of Defense*, 188–230.

11. HQ, 82nd Airborne Division, Historical Summary, 1968–1975, Research Material box 54, WMM82; Maj. Gen. Roscoe Robinson, "Deployment Alert," May 23, 1978, ROAD box 5, WMM82; HQ, 82nd Airborne Division, FY1979 Historical Summary, Research Material box 55, WMM82; John E. Valliere, "Disaster at Desert One: Catalyst for Change," *Parameters* 22, no. 1 (Autumn 1992): 69–82.

12. David W. Hogan Jr., *Raiders or Elite Infantry? The Changing Role of the US Army Rangers from Dieppe to Grenada* (Westport, CT: Greenwood, 1992), 36–51; Department of the Army, *Ranger Unit Operations*, Field Manual 7-85, 1987.

13. John K. Cooley, "US Rapid Strike Force: How to Get There First with the Most," *Christian Science Monitor*, April 11, 1980; US House of Representatives, *House Committee on Appropriations, Department of Defense Appropriations for 1983*, 97th Congress, 2nd Session, 306; Richard Halloran, "Pentagon Activates Strike Force; Effectiveness Believed Years Off," *New York Times*, February 19, 1980.

14. Stansfield Turner, "Towards a New Defense Strategy," *New York Times*, May 10, 1981; US House of Representatives, DoD Appropriations Hearing for 1983, 309.

15. Richard Stewart, *Operation Urgent Fury: The Invasion of Grenada, October 1983* (Washington, DC: GPO, 2009), 18–22, 26, 37; R. Cody Phillips, *Operation Just Cause: The Incursion into Panama*, Center for Military History Publication 70-85-1 (Washington, DC: GPO, 2004), 32; Executive Summary, XVIII Airborne Corps, June 24, 1991, Gulf War, box 43, WMM82; Scales, *Certain Victory*, 49–51; John L. Romjue, *American Army Doctrine for the Post–Cold War* (Washington, DC: CMH, 1997), 90, 101; Steve Krippel and Chris Riccie, "The Stryker Brigade Combat Team: America's Early Entry Force," *Infantry Journal*, July–September 2014, 26–29.

16. R. F. M. Williams, "The Development of Airfield Seizure Operations in the United States Army," *Military Review*, November 18, 2021; Richard W. Stewart, *The United States Army in Afghanistan, Operation Enduring Freedom, October 2001–March 2002* (Washington, DC: CMH, 2004), 14; Robert W. Jones Jr., "The Jump at Objective Serpent: 3/75th Rangers in Iraq," *Veritas* 1, no. 1 (2005): 52–54; Peter R. Mansoor, *Surge: My Journey with General David Petraeus and the Remaking of the Iraq War* (New Haven, CT: Yale University Press, 2013), 70.

17. Jim Garamone, "Carter Calls Bragg Troops 'Tip of Spear' for New Strategic Era," *DoD News*, July 14, 2015, https://www.army.mil/article/152226/carter_calls_bragg_troops_tip_of_spear_for_new_strategic_era; NDC Conference Report, "The Future of Airborne Forces in NATO," Research Division, NATO Defense College, Rome, July 2013, 3; Donna Miles, "From Haiti to Afghanistan, 82nd Shows Flexibility," *American Forces Press Service*, April 30,

2010, https://www.globalsecurity.org/military/library/news/2010/04/mil-100430-afps02.htm.

18. Matthew Cox, "Emergency Paratrooper Deployment Is First for New Army Response Force," *Military.com*, January 3, 2020, https://www.military.com/daily-news/2020/01/02/emergency-army-deployment-first-new-paratrooper-response-force.html; Jay Price, "82nd Airborne Division Celebrates 100 Years," NPR, August 24, 2017, https://www.npr.org/2017/08/24/545757890/82nd-airborne-division-celebrates-its-100th-anniversary.

19. Ryan Pickrell, "The US Military Moved 1,600 Soldiers into Positions outside the Nation's Capital and Has Them on Alert to Respond to Protests If Necessary," Business Insider, June 2, 2020, https://www.businessinsider.com/pentagon-1600-troops-on-alert-outside-dc-for-protest-response-2020-6?op=1; Jack Murphy, "New Details about the 82nd Military Deployment to D.C.," *Connecting Vets*, June 11, 2020, https://www.audacy.com/connectingvets/articles/new-details-about-the-82nd-military-deployment-to-dc; Jack Murphy, "82nd Paratroopers Forward Deployed near Washington DC," *Connecting Vets*, June 2, 2020, https://www.audacy.com/connectingvets/articles/82nd-paratroopers-forward-deployed-near-washington-dc; Robert Burns, Matthew Lee, and Ellen Knickmeyer, "US Sending 3K Troops for Partial Afghan Embassy Evacuation," AP, August 12, 2021, https://apnews.com/article/afghanistan-us-troops-embassy-kabul-355c48ec08fb7eb75e1e279e99c3dabf.

20. Dan Fastenberg and Bryan Woolston, "U.S. Troops Prepare for Deployment to Eastern Europe from Fort Bragg," Reuters, February 3, 2022, https://www.reuters.com/world/us-troops-prepare-deployment-eastern-europe-fort-bragg-2022-02-03/; Rachel Riley, "Fort Bragg Soldiers Deploy to Support NATO in Europe amid Russian Threat to Ukraine," *Fayetteville (NC) Observer*, February 3, 2022, https://www.fayobserver.com/story/news/2022/02/03/fort-bragg-soldiers-deploy-europe-russia-presence-ukraine-border-putin-us-military/6646815001/.

21. Bernard Loeffke to John Foss, August 8, 1988, box 2A, John W. Foss Papers, USAHEC; Timothy S. Muchmore, "Redefining Roles and Missions and Restructuring Forces of the Army and Marine Corps" (MA thesis, Naval War College, 1991), 9, Army Roles and Missions Collection, box 2B, USAHEC; George W. Bush, Presidential Proclamation, "National Airborne Day, 2002," August 14, 2002; S.Res.235–111th Congress (2009–2010): "A Resolution Designating August 16, 2009, as 'National Airborne Day,'" legislation, August 3, 2009, 1st Session, 111th Congress; "West Point's Army-Navy Game Uniforms to Honor 82nd Airborne," *Stars and Stripes*, December 6, 2016, https://www.stripes.com/sports/west-point-s-army-navy-game-uniforms-to-honor-82nd-airborne-1.442946; J. D. Simkins, "Army to Honor 1st Cavalry Division with New Unis against Rival Navy," *Military Times*, December 6, 2019, https://www.militarytimes.com/off-duty/military-culture/2019/12/06/army-uniforms-honor-1st-cavalry-division-for-navy-rivalry-game/.

22. Steve Beynon, "New Army 11th Airborne Division Gets Stand Up Date, Force Outline," Military.com, May 18, 2022, https://www.military.com/daily-news/2022/05/18/new-army-11th-airborne-division-gets-stand-date-force-outline.html. For the origins of "hooah" see Sarah Sicard,

"The Mysterious Origins of 'HOOAH,' the Army's Beloved Battle Cry," Task & Purpose, October 5, 2017, https://taskandpurpose.com/history/mysterious-origins-hooah-armys-beloved-battle-cry/.

23. Meghann Myers, "Earning It: A Complete History of Army Berets and Who's Allowed to Wear Them," *Army Times*, November 19, 2017, https://www.armytimes.com/news/your-army/2017/11/19/earning-it-a-complete-history-of-army-berets-and-whos-allowed-to-wear-them/; Correspondence about Beret, 1979, Reference Material, Training, box 32, WMM82; "Chief of Staff Hints He'd Wear Berets," *Army Times*, October 1, 1979.

24. DA, Change 4, Army Regulation 600–70, January 24, 1950; Thomas Hendrix, "The Parachutist Badge—68 Proud Years," www.army.mil, March 2, 2009, https://www.army.mil/article/17655/the_parachutist_badge_68_proud_years; DA, Army Regulation 600–8–22, *Military Awards* (2019), 117–19; Beaumont, *Military Elites*, 2–3. For more on rites of passage see Arnold Van Gennep, *The Rites of Passage*, trans. M. B. Visedom and G. L. Caffe (Chicago: University of Chicago Press, 1960).

25. Price, "82nd Airborne Division Celebrates 100 Years."

26. Brief history of the 82nd Airborne Division War Memorial Museum, Reference Material, Museum Info, box 57, WMM82; John W. Aarsen, "The 82d Airborne Division War Memorial Museum," *Army History*, no. 128 (Summer 2023): 20–23; Davis Winkie, "82nd Airborne's Special Relationship with D-Day Village Endures Virtually amid Pandemic," *Army Times*, December 16, 2020, https://www.armytimes.com/news/your-army/2020/12/15/82nd-airbornes-special-relationship-with-d-day-village-endures-virtually-amid-pandemic/.

27. Davis Winkie, "Why the 82nd Airborne Is Directing Airfield Security for Afghanistan Evacuation," *Army Times*, August 17, 2021, https://www.militarytimes.com/flashpoints/afghanistan/2021/08/17/why-the-82nd-airborne-is-directing-airfield-security-for-afghanistan-evacuation/; "Last Boots in Afghanistan on Display in the Museum," *Call to Duty: Newsletter of the Army Historical Foundation and the National Museum of the United States Army* 17, no. 3 (September 2022): 8; Jeff School, "Why a 2-Star General Was the Last American Service Member to Leave Afghanistan," Task & Purpose, August 31, 2021, https://taskandpurpose.com/news/army-general-last-soldier-leave-afghanistan/.

28. Haley Britzky, "The True Story of How Army Paratroopers Traded Dip for a Toyota Gun Truck Used to Secure the Kabul Airport," Task & Purpose, October 6, 2021, https://taskandpurpose.com/news/army-paratroopers-toyota-technical-kabul-airport/; Alex Horton and Dan Lamothe, "Inside the Afghanistan Airlift: Split-Second Decisions, Relentless Chaos Drove Historic Military Mission," *Washington Post*, September 27, 2021, https://www.washingtonpost.com/national-security/2021/09/27/afghanistan-airlift-inside-military-mission/.

29. Wyatt Olson, "US Army Alaska to Be Reflagged as Airborne Division amid Surge in Troop Suicides," *Stars and Stripes*, May 6, 2022, https://www.stripes.com/theaters/us/2022-05-05/us-army-alaska-11th-airborne-division-arctic-strategy-5910999.html; R. F. M. Williams, "Bring Back the Sightseeing Sixth: The Case for an Arctic Division," Modern War Institute, December 14, 2021, https://mwi.usma.edu/bring-back-the-sightseeing-sixth-the-case-for-an-arctic-division/; Stephen Spielberg and Tom Hanks, producers, *Band of Brothers*

(HBO Entertainment, 2001); Randall Wallace, dir., *We Were Soldiers* (Paramount Pictures, 2022); Ridley Scott, dir., *Black Hawk Down* (Sony Pictures, 2001).

30. Devore, "When Failure Thrives," 58–59.

31. Winski and McCaffery quoted in Kyle Jahner, "Does the Army Even Need Airborne?," *Army Times*, February 29, 2016, https://www.armytimes.com/news/your-army/2016/02/29/does-the-army-need-airborne/; Logan Nye, "How the 'Little Groups of Paratroopers' Became Airborne Legends," We Are the Mighty, April 2, 2018, https://www.wearethemighty.com/articles/how-the-little-groups-of-paratroopers-became-airborne-legends/.

32. Walter A. Schrepel, "Paras and Centurions: Lessons Learned from the Battle of Algiers," *Peace and Conflict: Journal of Peace Psychology* 11, no. 1 (July 2005): 71–89, https://doi.org/10.1207/s15327949pac1101_9; Beaumont, *Military Elites*, 111–12; Rod Thornton, "Getting It Wrong: The Crucial Mistakes Made in the Early Stages of the British Army's Deployment to Northern Ireland (August 1969 to March 1972)," *Journal of Strategic Studies* 30, no. 1 (February 1, 2007): 73–107, https://doi.org/10.1080/01402390701210848; Winslow, *Canadian Airborne Regiment in Somalia*.

Bibliography

Abbreviations

BHL	Bentley Historical Library, Ann Arbor, MI
CARL	Combined Arms Research Library, Fort Leavenworth, KS
CGSC	Command and General Staff College, Fort Leavenworth, KS
CMH	US Army Center of Military History, Fort McNair, Washington, DC
CSI	Combat Studies Institute, Fort Leavenworth, KS
DA	Department of the Army
DDEL	Dwight D. Eisenhower Library, Abilene, KS
DoD	Department of Defense
DRL	Donovan Research Library, Fort Moore, GA
DTIC	Defense Technical Information Center, Fort Belvoir, VA
FDRL	Franklin D. Roosevelt Library, Hyde Park, NY
FRUS	*Foreign Relations of the Unites States*, US Department of State, Washington, DC
GCMRL	George C. Marshall Research Library, Lexington, VA
GPO	US Government Printing Office
HSTL	Harry S. Truman Library, Independence, MO
JFKL	John F. Kennedy Library, Boston
JFK PP	John F. Kennedy Personal Papers
LBJL	Lyndon B. Johnson Library, Austin, TX
LOC	Library of Congress, Washington, DC
NARA	National Archives and Records Administration, College Park, MD
NDU	National Defense University, Fort McNair, Washington, DC
NLAU	Nicholson Library, Anderson University, IN
RG	Record Group
TTUVA	Vietnam Archive, Texas Tech University, Lubbock
USAHEC	United States Army Heritage and Education Center, Carlisle, PA
USMAA	United States Military Academy Archives, West Point, NY
WMM82	82nd Airborne Division War Memorial and Museum, Fort Liberty, NC

Archival Sources

American Presidency Project, University of California at Santa Barbara
Bentley Historical Library, University of Michigan, Ann Arbor
Combined Arms Research Library, Archive, Fort Leavenworth, KS
Donovan Research Library, Fort Benning, GA
Dwight D. Eisenhower Library, Abilene, KS

82nd Airborne Division War Memorial and Museum, Fort Liberty, NC
George C. Marshall Research Library, Lexington, VA
Harry S. Truman Library, Independence, MO
John F. Kennedy Presidential Library, Boston
Library of Congress, Washington, DC
Lyndon B. Johnson Presidential Library, Austin, TX
Mahn Center for Archives and Special Collections, Ohio University, Athens
Mudd Manuscript Library, Princeton, NJ
National Archives and Records Administration, College Park, MD
National Defense University, Washington, DC
Nicholson Library Special Collections, Anderson University, Anderson, IN
Rutgers Oral History Archives, Rutgers University, New Brunswick, NJ
United States Army Heritage and Education Center, Carlisle, PA
United States Military Academy, Archives and Manuscripts Division, West
    Point, NY
Vietnam Center and Sam Johnson Vietnam Archive, Texas Tech University,
    Lubbock
Yale University Library, Manuscripts and Archives, New Haven, CT

Other Sources

Aarsen, John W. "The 82d Airborne Division War Memorial Museum." *Army History*, no. 128 (Summer 2023): 20-23.
Air Force Magazine. "The Army's Bird in Hand." June 1956.
Alexander, Mark J., and John Sparry. *Jump Commander: In Combat with the 505th and 508th Parachute Infantry Regiments, 82nd Airborne Division in World War II.* Philadelphia: Casemate, 2018.
Allen, Matthew. *Military Helicopter Doctrines of the Major Powers 1945–1992: Making Decisions about Air-Land Warfare.* Westport, CT: Greenwood, 1993.
Ambrose, Stephen E. *Band of Brothers: E Company, 506th Regiment, 101st Airborne from Normandy to Hitler's Eagle's Nest.* New York: Simon & Schuster, 2001.
Andrews, Marshall. "Our New Army." *Washington Post*, May 9, 1942.
Angress, Werner T. *Witness to the Storm: A Jewish Journey from Nazi Berlin to the 82nd Airborne, 1920–1945.* Bloomington: Indiana University Press, 2012.
Aran, Gideon. "Parachuting." *American Journal of Sociology* 80, no. 1 (July 1974): 150. https://doi.org/10.1086/225764.
Army and Navy Bulletin. "Gavin Says Air Mechanization Will Win Wars of the Future." April 5, 1947.
Army Information Digest. "STRAC—Mobile Striking Force in Readiness." Vol. 13, no. 6 (August 1958): 16-17.
Army–Navy–Air Force Journal and Register. "Army Looks to Sage Brush's SkyCav Operation as Basis of Permanent Unit." December 10, 1955.
———. "Army to Retain Gen. Ridgway." January 22, 1955.
———. "New Divisional Organization." February 12, 1955.
Army Times. "Chief of Staff Hints He'd Wear Berets." October 1, 1979.
Astor, Gerald. *Battling Buzzards: The Odyssey of the 517th Parachute Regimental Combat Team 1942-1945.* New York: Dell, 1993.

Atkinson, Rick. *An Army at Dawn*. New York: Henry Holt, 2002.

Bacevich, Andrew J. "The Paradox of Professionalism: Eisenhower, Ridgway, and the Challenge to Civilian Control, 1953–1955." *Journal of Military History* 61, no. 2 (April 1997): 303–33. https://www.proquest.com/scholarly-journals/paradox-professionalism-eisenhower-ridgway/docview/1296645910/se-2.

——. *The Pentomic Era: The US Army between Korea and Vietnam*. Washington, DC: National Defense University, 1986.

Baldwin, Hanson W. "The Cavalry Charges On." *New York Times*, July 2, 1950.

——. "Gavin Explains Decision; Warns of 'Another Korea.'" *New York Times*, January 12, 1958.

——. "Infantry Adjusts Role." *New York Times*, November 23, 1953.

——. "Need of Training Revealed in Korea." *New York Times*, November 3, 1950.

——. "The New Army Corps." *New York Times*, May 24, 1958.

——. "Ridgway to the Rescue." *New York Times*, November 23, 1954.

——. "Ridgway vs. Eisenhower: A Review of the Apparent Contradiction in Their Remarks on Manpower Slashes." *New York Times*, January 24, 1956.

——. "Strategy in Vietnam." *New York Times*, September 21, 1963.

Ball, Lamar Q. "Today's Paratrooper Scorns Fatalism Fad." *Atlanta Constitution*, November 27, 1942.

Bando, Mark. *101st Airborne: The Screaming Eagles in World War II*. St. Paul, MN: Zenith, 2007.

Barlow, Jeffrey C. *Revolt of the Admirals: The Fight for Naval Aviation, 1945–1950*. Washington, DC: Naval Historical Center, 1993.

Barno, David, and Nora Benashel. *Adaptation under Fire: How Militaries Change in Wartime*. New York: Oxford University Press, 2020.

Beaumont, Roger A. "Airborne: Life Cycle of a Military Subculture." *Military Review* 51, no. 6 (June 1971): 52–61.

——. *Military Elites: Special Fighting Units in the Modern World*. New York: Bobbs-Merrill, 1974.

Bergerson, Frederic A. *The Army Gets an Air Force: Tactics of Insurgent Bureaucratic Politics*. Baltimore: Johns Hopkins University Press, 1980.

Bergman, David, Marie Gustafsson Senden, and Erik Berntson. "Preparing to Lead in Combat: Development of Leadership Self-Efficacy by Static-Line Parachuting." *Military Psychology* 31, no. 6 (October 2019): 481–89. https://doi.org/10.1080/08995605.2019.1670583.

Beynon, Steve. "New Army 11th Airborne Division Gets Stand Up Date, Force Outline." Military.com, May 18, 2022. https://www.military.com/daily-news/2022/05/18/new-army-11th-airborne-division-gets-stand-date-force-outline.html.

Biddle, Tami Davis. *Rhetoric and Reality in Air Warfare: The Evolution of British and American Ideas about Strategic Bombing, 1914–1945*. Princeton, NJ: Princeton University Press, 2002.

Biggs, Bradley. *Gavin: A Biography of General James M. Gavin*. Hamden, CT: Archon Books, 1980.

Birtle, Andrew. *US Army Counterinsurgency and Contingency Operations Doctrine, 1942–1976*. Washington, DC: CMH, 2006.

Blair, Clay. *The Forgotten War*. New York: Times Books, 1987.

———. *Ridgway's Paratroopers: The American Airborne in World War II*. Garden City, NY: Dial, 1985.

Blair, William M. "Wilson Restricts Army on Missiles and Air Program." *New York Times*, November 27, 1956.

Blumenson, Martin. *Salerno to Cassino: The Mediterranean Theater of Operations*. Washington, DC: GPO, 1969.

Booth, T. Michael, and Duncan Spencer. *Paratrooper: The Life of Gen. James M. Gavin*. New York: Simon & Schuster, 1994.

Bowers, Roy L. *Tactical Airlift*. The United States Air Force in Southeast Asia. Washington, DC: Office of Air Force History, 1983.

Boyle, Kevin. *The UAW and the Heyday of American Liberalism, 1945–1968*. Ithaca, NY: Cornell University Press, 1995.

Boyne, Walter J. *How the Helicopter Changed Modern Warfare*. Gretna, LA: Pelican, 2011.

Bradley, Francis X. "The Fallacy of Dual Capability." *Army* 10 (October 1959): 18–19.

Bradley, Omar N. "Creating a Sound Military Force." *Military Review* 24, no. 2 (May 1949): 3–6.

———. *A Soldier's Story*. New York: Henry Holt, 1951.

Bragg, James W. *Development of the Corporal: The Embryo of the Army Missile Program*. Redstone Arsenal, AL: Army Ballistic Missile Agency, 1961.

Breuer, William B. *Drop Zone Sicily: Allied Airborne Strike, July 1943*. Novato, CA: Presidio, 1983.

———. *Geronimo! American Paratroopers in World War II*. New York: St. Martin's, 1992.

Britzky, Haley. "The True Story of How Army Paratroopers Traded Dip for a Toyota Gun Truck Used to Secure the Kabul Airport." Task & Purpose, October 6, 2021. https://taskandpurpose.com/news/army-paratroopers-toyota-technical-kabul-airport/.

Brodie, Bernard, ed. *The Absolute Weapon*. New York: Harcourt Brace, 1946.

Brooks, Drew. "Obscure Fight in Dominican Republic Taught 82nd Airborne Urban Warfare Lessons." *Fayetteville Observer*, April 25, 2015. https://www.fayobserver.com/story/news/military/2015/04/25/obscure-fight-in-dominican-republic/22221364007/.

Browne, Malcom W. *The New Face of War*. Rev. ed. Indianapolis: Bobbs-Merrill, 1968.

Brumfield, Hunter M. "Swarmer—a Pattern for Airborne Assault." *Army Information Digest* 5, no. 7 (July 1950): 13–22.

Buccino, Joe. "Ike vs. Ridgway: Lessons for Today from the Philosophical Battle between Two of America's Greatest Military Leaders." *Modern War Institute*, April 14, 2020. https://mwi.usma.edu/ike-vs-ridgway-lessons-today-philosophical-battle-two-americas-greatest-military-leaders/.

Burgett, Donald R. *Currahee! A Screaming Eagle at Normandy*. Novato, CA: Presidio, 1967.

Burke, Eric Michael. *Soldiers from Experience: The Forging of Sherman's Fifteenth Army Corps, 1862–63*. Baton Rouge: Louisiana State University Press, 2022.

Burns, Dwayne, with Leland Burns. *Jump into the Valley of the Shadow: The World War II Memories of a Paratrooper in the 508th P.I.R., 82nd Airborne Division.* Philadelphia: Casemate, 2006.

Burns, Robert, Matthew Lee, and Ellen Knickmeyer. "US Sending 3K Troops for Partial Afghan Embassy Evacuation." AP, August 12, 2021. https://apnews.com/article/afghanistan-us-troops-embassy-kabul-355c48ec08f b7eb75e1e279e99c3dabf.

Busch, Briton Cooper. *Bunker Hill to Bastogne: Elite Forces and American Society.* Dulles, VA: Potomac Books, 2006.

Cameron, Craig M. *American Samurai: Myth, Imagination, and the Conduct of Battle in the First Marine Division, 1941–1951.* New York: Cambridge University Press, 1994.

Campbell, Kenneth. "Pace Urges Army Built on Science." *New York Times,* June 7, 1950.

Carland, John M. *Combat Operations: Stemming the Tide, May 1965 to October 1966.* Washington, DC: CMH, 2000.

Carter, Donald A. "Eisenhower versus the Generals." *Journal of Military History* 71, no. 4 (October 2007): 1169–99. https://doi.org/10.1353/jmh.2007. a222498.

———. *Forging the Shield: The U.S. Army in Europe, 1951–1952.* Washington, DC: CMH, 2015.

———. *The US Army before Vietnam.* Washington, DC: CMH, 2015.

Chaco, Tania M. "Why Did They Fight? American Airborne Units in World War II." *Defence Studies* 1, no. 3 (Autumn 2001): 59–94. https://doi. org/10.1080/714000045.

Cheng, Christopher C. S. *Air Mobility: The Development of a Doctrine.* Westport, CT: Praeger, 1994.

Clark, Lloyd. *Crossing the Rhine: Breaking into Nazi Germany, 1944 and 1945—the Greatest Airborne Battles in History.* New York: Atlantic Monthly, 2008.

Clark, Mark. *Calculated Risk.* New York: Harper Collins, 1950.

Clarke, Bruce. C. "Abe." *Armor* 84, no. 1 (January–February 1975): 17.

Cockerham, William C. "Selective Socialization: Airborne Training as a Status Passage." *Journal of Political and Military Sociology* 1 (Fall 1973): 215–29. http://www.jstor.org/stable/45293603.

Cockerham, William C., and Lawrence E. Cohen. "Volunteering for Foreign Combat Missions: An Attitudinal Study of U.S. Army Paratroopers." *Pacific Sociological Review* 24, no. 3 (July 1981): 329–54. https://doi. org/10.2307/1388810.

Coffman, Edward M. *The Regulars: The American Army, 1898–1941.* Cambridge, MA: Harvard University Press, 2004.

Cole, Alice C., ed. *The Department of Defense: Documents on Establishment and Organization, 1947–1978.* Washington, DC: Office of the Secretary of Defense Historical Office, 1978.

Coleman, J. D. *Pleiku: The Dawn of Helicopter Warfare in Vietnam.* New York: St. Martin's, 1988.

Collins, Arthur S. "The Other Side of the Atom." *Army* 10 (November 1959): 18–19.

Compton, Lynn "Buck," with Marcus Brotherton. *Call of Duty: My Life before, during, and after the Band of Brothers*. New York: Berkley Caliber, 2008.

Cooley, John K. "US Rapid Strike Force: How to Get There First with the Most." *Christian Science Monitor*, April 11, 1980.

Cox, Matthew. "Emergency Paratrooper Deployment Is First for New Army Response Force." Military.com, January 3, 2020. https://www.military.com/daily-news/2020/01/02/emergency-army-deployment-first-new-paratrooper-response-force.html.

Crane, Conrad C. "Matthew Ridgway and the Value of Persistent Dissent." *Parameters* 51, no. 2 (Summer 2021): 7–18. https:/doi.org/10.55540/0031-1723.3064.

———. "Phase IV Operations: Where Wars Are Really Won." *Military Review*, May–June 2005, 11–20. https://www.armyupress.army.mil/Portals/7/PDF-UA-docs/Crane-2006-UA.pdf.

Culp, W. W. "Resident Courses of Instruction." *Military Review* 36 (May 1956): 15–21.

Curatola, John M. *Bigger Bombs for a Brighter Tomorrow: The Strategic Air Command and American War Plans at the Dawn of the Atomic Age, 1945–1950*. Jefferson, NC: McFarland, 2015.

Daddis, Gregory A. *No Sure Victory: Measuring U.S. Army Effectiveness and Progress in the Vietnam War*. New York: Oxford University Press, 2011.

———. *Westmoreland's War: Reassessing American Strategy in Vietnam*. New York: Oxford University Press, 2015.

Dastrup, Boyd L. *The US Army Command and General Staff College: A Centennial History*. Manhattan, KS: Sunflower University Press, 1982.

Davis, Robert T. *The Challenge of Adaptation: The US Army in the Aftermath of Conflict, 1953–2000*. Fort Leavenworth, KS: CSI Press, 2008.

Dawson, W. Forrest, ed. *Saga of the All American*. Atlanta: Albert Love, 1946.

Dealie, Melvin G. *Always at War: Organizational Culture in Strategic Air Command, 1946–62*. Annapolis, MD: Naval Institute Press, 2018.

Deam, Donald L. *General Toothpick: The WWII Memoirs of 1st Sgt. Donald L. Deam*. Fort Campbell, KY: 101st Airborne Division, 2008.

Department of the Army. *Airmobile Operations*. Field Manual 57–35, 1960.

———. *Army Air Defense Operations*. Army Regulation 525–50, 1956.

———. *Army Missiles*. Army Regulation 525–20, 1956.

———. *Army Missiles and Rockets*. DA Pamphlet 355–13, May 1958.

———. *Army Mobility Concept*. Washington, DC: GPO, December 1963.

———. *Army Transport Aviation Combat Operations*. Field Manual 57–35, 1958.

———. *Davy Crockett Weapons System in Infantry and Armor Units*. Field Manual 23–20, 1961.

———. *Field Service Regulations: Operations*. Field Manual 100–5, each edition, 1949–1993.

———. *Field Service Regulations: Operations*. Field Manual 3–0, each edition, 2001–2022.

———. *Military Awards*. Army Regulation 600–8–22, 2019.

———. *Ranger Unit Operations*. Field Manual 7–85, 1987.

Department of Defense, *Annual Report for Fiscal Year 1964.* Washington, DC: GPO, 1965.

———. *Annual Report of the Secretary of Defense.* FY 1960.

———. *Minutes of Meetings of the Combined Chiefs of Staff, Post-Arcadia.* Vol. 1. Washington, DC: Joint History Office, 2003.

———. *Semiannual Report of the Secretaries of the Army, Navy, and Air Force.* Multiple eds. Washington, DC: GPO, 1954–1961.

Department of State. *Foreign Relations of the United States.* Multiple volumes for 1949–1968. Washington, DC: GPO, 1975–2003.

DePuy, William E. "The Case for Dual Capability." *Army* 10 (January 1960): 32–41.

Devers, Jacob L. "Air Transportability of the Infantry Division." *Military Review* 24, no. 1 (April 1949): 14–18.

Devlin, Gerald M. *Paratrooper! The Saga of the U.S. Army and Marine Parachute and Glider Combat Troops during World War II.* New York: St. Martin's, 1986.

———. *Silent Wings: The Saga of the U.S. Army and Marine Combat Glider Pilots during World War II.* New York: St. Martin's, 1985.

DeVore, Marc R. *When Failure Thrives: Institutions and the Evolution of Postwar Airborne Forces.* Fort Leavenworth, KS: Army Press, 2015.

Disney, Paul. "Armor in Atomic Warfare." *Armor* 63 (May–June 1953): 30–31.

Dobbs, Michael. *One Minute to Midnight: Kennedy, Khrushchev, and Castro on the Brink of Nuclear War.* New York: Vintage Books, 2008.

Donahue, Christopher. "We Do Not 'Heavy Breathe' in the 82nd Airborne Division." *18th Airborne Corps Podcast,* episode 37, June 7, 2021. Produced by Joe Buccino. https://thedoomsdayclock.podbean.com/e/episode-37-we-do-not-heavy-breathe-in-this-division-a-discussion-with-maj-gen-chris-donahue/.

Donnelly, William M. "Bilko's Army: A Crisis in Command?" *Journal of Military History* 75, no. 4 (October 2011): 1183–1215. https://search.ebscohost.com/login.aspx?direct=true&db=smf&AN=66649429&site=ehost-live.

Drury, Allen. "Gavin's Quitting Will Be Studied by Senate Unit." *New York Times,* January 6, 1958.

Duranty, Walter. "Soviets Initiate Parachute Attack." *New York Times,* September 16, 1935.

Dvorchak, Robert J. *Golden Brigade: The Untold Story of the 82nd Airborne in Vietnam and Beyond.* Indianapolis: IBJ, 2020.

Eddleman, C. D. "The Pentomic Reorganization—a Status Report." *Army Information Digest* 13, no. 9 (September 1958): 3–4.

Eisenhower, Dwight D. *Crusade in Europe.* New York: Doubleday, 1948.

———. *Mandate for Change.* New York: Doubleday, 1963.

———. *The Papers of Dwight D. Eisenhower: The War Years.* 5 vols. Edited by Alfred D. Chandler. Baltimore: Johns Hopkins University Press, 1970.

Eisenhower, John S. D. *The Bitter Woods: The Battle of the Bulge.* New York: Da Capo, 1969.

Elliot, David C. "Project Vista and Nuclear Weapons in Europe." *International Security* 11, no. 1 (1986): 163–83. https://doi.org/10.2307/2538879.

Ellis, John T. *The Airborne Command and Center*. Army Ground Forces Historical Section Study No. 25, 1946.

English, Allan D. *Understanding Military Culture: A Canadian Perspective*. Montreal: McGill-Queen's University Press, 2004.

English, John A., and Bryce I. Gudmundsson. *On Infantry*. Westport, CT: Praeger, 1994.

Erikson, Jane. "US Surgeon Lindsey, Korean War Vet, Dies." *Arizona Daily Star*, January 5, 2006.

Evans, Michael. "The Primacy of Doctrine: The United States Army and Military Innovation and Reform, 1945–1995." Army Occasional Paper No. 1. Washington, DC, 1996.

Fall, Bernard B. *Hell in a Very Small Place: The Siege of Dien Bien Phu*. Philadelphia: Lippincott, 1967.

———. *Street without Joy: The French Debacle in Indochina*. Guilford, CT: Stackpole Books, 2018.

Fastenberg, Dan, and Bryan Woolston. "U.S. Troops Prepare for Deployment to Eastern Europe from Fort Bragg." Reuters, February 3, 2022. https://www.reuters.com/world/us-troops-prepare-deployment-eastern-europe-fort-bragg-2022-02-03/.

Fauntleroy, Barbara Gavin. *The General and His Daughter: The Wartime Letters of General James M. Gavin to His Daughter Barbara*. New York: Fordham University Press, 2007.

Fautua, David T. "The 'Long Pull' Army: NSC 68, the Korean War, and the Creation of the Cold War U.S. Army." *Journal of Military History* 61, no. 1 (January 1997): 93–120. https://doi.org/10.2307/2953916.

Fenelon, James M. *Four Hours of Fury: The Untold Story of World War II's Largest Airborne Invasion and the Final Push into Nazi Germany*. New York: Scribner, 2019.

Ferrell, Robert H., ed. *The Diary of James C. Hagerty: Eisenhower in Mid-course, 1954–1955*. Bloomington: Indiana University Press, 1983.

Finlayson, Kenneth. "Lieutenant General William P. Yarborough." *Veritas* 2, no. 2 (2006): 42–49. https://arsof-history.org/articles/pdf/v2n2_yarborough.pdf.

Fisher, Ernest F. "Evolution of U.S. Airborne Doctrine." *Military Review*, May 1966, 71–77.

Fitzgerald, David. *Learning to Forget: US Army Counterinsurgency Doctrine and Practice from Vietnam to Iraq*. Stanford, CA: Stanford University Press, 2014.

Flanagan, E. M. *Airborne: A Combat History of American Airborne Forces*. New York: Presidio, 2002.

Fowler, Randall. *More Than a Doctrine: The Eisenhower Era in the Middle East*. Lincoln, NE: Potomac Books, 2018.

Freedman, Lawrence, and Jeffrey Michaels. *The Evolution of Nuclear Strategy*. 4th ed. London: Palgrave Macmillan, 2019.

Furgurson, Ernest B. *Westmoreland: The Inevitable General*. Boston: Little, Brown, 1968.

Futrell, Robert Frank. *Ideas, Concepts, Doctrine.* Vol. 1, *Basic Thinking in the United States Air Force, 1907–1960.* Maxwell Air Force Base, AL: Air University Press, 2004.

Gabel, Christopher R. *The U.S. Army GHQ Maneuvers of 1941.* Washington, DC: GPO, 1992.

Gabel, Kurt. *The Making of a Paratrooper: Airborne Training and Combat in World War II.* Lawrence: University Press of Kansas, 1990.

Gaddis, John Lewis. *The Cold War: A New History.* New York: Penguin Books, 2007.

———. *Strategies of Containment: A Critical Appraisal of American National Security Policy during the Cold War.* New York: Oxford University Press, 2005.

Galvin, John R. *Fighting the Cold War: A Soldier's Memoir.* Lexington: University Press of Kentucky, 2015.

Garamone, Jim. "Carter Calls Bragg Troops 'Tip of Spear' for New Strategic Era." *DoD News,* July 14, 2015. https://www.army.mil/article/152226/carter_calls_bragg_troops_tip_of_spear_for_new_strategic_era.

Garland, Albert N., Howard M. Smyth, and Martin Blumenson. *Sicily and the Surrender of Italy: The Mediterranean Theater of Operations.* Washington, DC: Department of the Army, 1965.

Gavin, James M. *Airborne Warfare.* Washington, DC: Infantry Journal Press, 1947.

———. "Arms Vigilance for Peace." *Ordnance* 39, no. 209 (March–April 1955): 716–19.

———. "Cavalry, and I Don't Mean Horses." *Harper's,* April 1954.

———. *Crisis Now.* New York: Random House, 1968.

———. "The Future of Airborne Operations." *Military Review* 27, no. 9 (December 1947): 3–8.

———. *On to Berlin: Battles of an Airborne Commander 1943–1946.* New York: Viking, 1978.

———. "The Tactical Use of the Atomic Bomb." *Combat Forces Journal* 1 (November 1950): 9–11.

———. *War and Peace in the Space Age.* New York: Harper, 1958.

———. "We Can Solve Our Technical Difficulties." *Army Combat Forces Journal,* November 1955, 64–65.

Geertz, Clifford. *The Interpretation of Cultures.* New York: Basic Books, 1973.

Gibbons, William Conrad. *The U.S. Government and the Vietnam War: Executive and Legislative Roles and Relationships, Part IV: July 1965–January 1968.* Princeton, NJ: Princeton University Press, 1995.

Giglio, James N. *The Presidency of John F. Kennedy.* Lawrence: University Press of Kansas, 2006.

Gladwell, Malcom. *The Bomber Mafia: A Dream, a Temptation, and the Longest Night of the Second World War.* New York: Little, Brown, 2021.

Goldstein, Richard. "Gen. H. H. Howze, 89, Dies; Proposed Copters as Cavalry." *New York Times,* December 18, 1998.

Goldstein, Sean. "Talks Barely Beat Invasion US Intervention in Haiti." *Baltimore Sun,* September 20, 1994.

Goodwins, Leslie, dir. *Parachute Battalion*. RKO Pictures, 1941.

Gordon, Michael R. "Heading Back to Iraq for Round 2." *New York Times*, March 1, 2004. https://www.nytimes.com/2004/03/01/international/worldspecial3/heading-back-to-iraq-for-round-2.html.

Grant, Rebecca. "Dien Bien Phu." *Air Force Magazine*, August 2004. https://www.airandspaceforces.com/article/0804dien/.

Greenberg, Lawrence M. *United States Army Unilateral and Coalition Operations in the 1965 Dominican Republic Intervention*. Washington, DC: CMH, 1987.

Greenfield, Kent R., Robert J. Palmer, and Bell I. Wiley. *The Organization of Ground Combat Troops*. U.S. Army in World War II. Washington, DC: Department of the Army, 1947.

Grenier, John. *The First Way of War: American War Making on the Frontier, 1607–1814*. New York: Cambridge University Press, 2010.

Guarnere, William, and Edward Heffron, with Robyn Post. *Brothers in Battle, Best of Friends: Two WWII Paratroopers from the Original Band of Brothers Tell Their Story*. New York: Berkley Caliber, 2007.

Gudmundsson, Bruce I. *Stormtroop Tactics: Innovation in the German Army, 1914–1918*. Westport, CT: Praeger, 1995.

Haggerty, James J. "No More Paratroops?" *Collier's*, March 18, 1955.

Halberstam, David. *The Best and the Brightest*. New York: Random House, 1969.

———. *The Fifties*. New York: Random House, 1993.

Halloran, Richard. "Pentagon Activates Strike Force; Effectiveness Believed Years Off." *New York Times*, February 19, 1980.

Halperin, Morton H. *Limited War in the Nuclear Age*. New York: John Wiley, 1963.

Hankins, Michael W. *Flying Camelot: The F-15, the F-16, and the Weaponization of Fighter Pilot Nostalgia*. Ithaca, NY: Cornell University Press, 2021.

Harclerode, Peter. *Wings of War: Airborne Warfare 1918–1945*. London: Cassell, 2005.

Harrington, Daniel F. *Berlin on the Brink: The Blockade, the Airlift, and the Early Cold War*. Lexington: University Press of Kentucky, 2012.

Hasik, Hayley Michael. "The Helicopter War: Unraveling the Myth and Memory of a Vietnam War Icon." PhD diss., University of Southern Mississippi, 2023.

Heilbrunn, Ott. *Conventional Warfare in the Nuclear Age*. New York: Praeger, 1965.

Hendrix, Thomas L. "The Parachutist Badge—68 Proud Years." army.mil., March 2, 2009. https://www.army.mil/article/17655/the_parachutist_badge_68_proud_years.

Herndon, James S., and Joseph O. Baylen. "Col. Philip R. Faymonville and the Red Army, 1934–43." *Slavic Review* 34, no. 3 (September 1975): 483–505. https://doi.org/10.2307/2495561.

Herring, George C., and Richard H. Immerman. "Eisenhower, Dulles, and Dienbienphu: 'The Day We Didn't Go to War' Revisited." *Journal of American History* 71, no. 2 (September 1984): 343–63. https://doi.org/10.2307/1901759.

Hewes, James E. *From Root to McNamara: Army Organization and Administration, 1900–1963*. Washington, DC: CMH, 1975.

Hitchcock, William I. *The Age of Eisenhower: America and the World in the 1950s.* New York: Simon & Schuster, 2019.

Hoffenaar, Jan, and Dieter Krüger, eds. *Blueprints for Battle: Planning for War in Central Europe, 1948–1968.* Lexington: University Press of Kentucky, 2012.

Hoffman, Frank G. *Mars Adapting: Military Change during War.* Annapolis, MD: Naval Institute Press, 2021.

Hoffman, Jon T., ed. *A History of Innovation: U.S. Army Adaptation in War and Peace.* Washington, DC: CMH, 2009.

Hogan, David W., Jr. *Raiders or Elite Infantry? The Changing Role of the US Army Rangers from Dieppe to Grenada.* Westport, CT: Greenwood, 1992.

Hogan, Michael J. *A Cross of Iron: Harry S. Truman and the Origins of the National Security State, 1945–1954.* Cambridge: Cambridge University Press, 2000.

Holm, Jeremy C. *When Angels Fall: From Toccoa to Tokyo.* Salt Lake City: Holm, 2019.

Holmes, Richard. *Acts of War: The Behavior of Men in Battle.* New York: Free Press, 1986.

Horton, Alex, and Dan Lamothe. "Inside the Afghanistan Airlift: Split-Second Decisions, Relentless Chaos Drove Historic Military Mission." *Washington Post*, September 27, 2021. https://www.washingtonpost.com/national-security/2021/09/27/afghanistan-airlift-inside-military-mission/.

House, Jonathan M. *A Military History of the Cold War, 1944–1962.* Norman: University of Oklahoma Press, 2012.

Howze, Hamilton H. *A Cavalryman's Story: Memoirs of a Twentieth-Century Army General.* Washington, DC: Smithsonian Institution, 1996.

——. "Combat Tactics for Tomorrow's Army." *Army* 8 (October 1958): 24–30.

——. "Tactical Employment of the Air Assault Division." *Army* 14, no. 2 (September 1963): 35–53.

Hull, Isabel V. *Absolute Destruction: Military Culture and the Practices of War in Imperial Germany.* Ithaca, NY: Cornell University Press, 2004.

Humphrey, Hubert. *The Education of a Public Man.* New York: Doubleday, 1976.

Huston, James A. *Out of the Blue: U.S. Army Airborne Operations in World War II.* West Lafayette, IN: Purdue University Press, 1972.

Hutcheson, Keith. *Air Mobility: The Evolution of Global Reach.* Vienna, VA: Point One, 1999.

Hutton, Carl I. "The Commandant's Column: An Air Fighting Army?" *Army Aviation Digest*, July 1955, 2.

Hyde, Justin. "Recalling an Awkward Phrase." *Detroit Free Press*, September 14, 2008.

Jahner, Kyle. "Does the Army Even Need Airborne?" *Army Times*, February 29, 2016. https://www.armytimes.com/news/your-army/2016/02/29/does-the-army-need-airborne/.

Johnson, E. R., and Lloyd S. Jones. *American Military Transport Aircraft since 1925.* Jefferson, NC: McFarland, 2013.

Jones, Robert W. "The Jump at Objective Serpent: 3/75th Rangers in Iraq." *Veritas* 1, no. 1 (2005): 52–54. https://arsof-history.org/articles/pdf/v1n1_objective_serpent.pdf.

Jussel, Paul C. "Intimidating the World: The United States Atomic Army, 1956–1960." PhD diss., Ohio State University, 2004.

Kaplan, Edward. *To Kill Nations: American Strategy in the Air-Atomic Age and the Rise of Mutually Assured Destruction*. Ithaca, NY: Cornell University, 2015.

Kellett, Donald T., and William Friedman. "Airborne on Paper Wings." *Infantry Journal* 62, no. 5 (May 1948): 9–14.

Kelly, Henry E. "Verbal Defense." *Military Review* 35, no. 7 (October 1955): 45–52.

Kennan, George F. *Realities of American Foreign Policy*. Princeton, NJ: Princeton University Press, 1954.

Kennedy, John F. *The Strategy of Peace*. Edited by Allan Nevins. New York: Harper & Row, 1960.

Kindsvatter, Peter S. *American Soldiers: Ground Combat in the World Wars, Korea, and Vietnam*. Lawrence: University Press of Kansas, 2003.

King, Anthony. *The Combat Soldier: Infantry Tactics and Cohesion in the Twentieth and Twenty-First Centuries*. New York: Oxford University Press, 2013.

Kingseed, Cole C. *Conversations with Major Dick Winters: Life Lessons from the Commander of the Band of Brothers*. New York: Berkley Caliber, 2014.

Kinnard, Douglas. "Civil-Military Relations: The President and the General." *Parameters* 15, no. 1 (1985): 19–29. https:/doi.org/10.55540/0031-1723.1387.

Kinnard, Harry W. O. "Activation to Combat—in 90 Days." *Army Information Digest* 21, no. 4 (April 1966): 25–31.

———. "Airmobility Revisited, Part 2." *Army Aviation Digest* 26, no. 7 (July 1980): 8–14.

———. "A Victory in the Ia Drang: The Triumph of a Concept." *Army* 17, no. 9 (September 1967): 71–91.

Kipp, Jacob W. "The Political Ballet of General Aleksandr Lebed." *Problems of Post-Communism* 43, no. 4 (July 1, 1996): 43–53. https://doi.org/10.1080/10758216.1996.11655688.

Kissinger, Henry. *Nuclear Weapons and Foreign Policy*. Garden City, NY: Doubleday, 1957.

Krepinevich, Andrew F. *The Army and Vietnam*. Baltimore: Johns Hopkins University Press, 1986.

Kretchik, Walter. *U.S. Army Doctrine: From the American Revolution to the War on Terror*. Lawrence: University Press of Kansas, 2011.

Kretchik, Walter, Robert F. Baumann, and John T. Fishel. *Invasion, Intervention, "Intervasion": A Concise History of the U.S. Army in Operation Uphold Democracy*. Fort Leavenworth, KS: CGSC Press, 1998.

Krippel, Steve, and Chris Riccie. "The Stryker Brigade Combat Team: America's Early Entry Force." *Infantry Journal*, July–September 2014, 26–29. https://www.moore.army.mil/infantry/magazine/issues/2014/Jul-Sep/Krippel.html.

Kuhn, W. A. "How Far Along Are We in Developing an Airborne Army?" *Military Review* 30, no. 1 (April 1950): 41–50.

Kurowski, Frank. *Jump into Hell: German Paratroopers in World War II*. Mechanicsburg, PA: Stackpole Books, 2010.

Lamb, David. "Copter Proves Itself as Vietnam Weapon." *Nashville Tennessean*, July 6, 1969.

Langrehr, Henry, and Jim DeFelice. *Whatever It Took: An American Paratrooper's Extraordinary Memoir of Escape, Survival, and Heroism in the Last Days of World War II*. New York: William Morrow, 2020.

"Last Boots in Afghanistan on Display in the Museum." *Call to Duty: Newsletter of the Army Historical Foundation and the National Museum of the United States Army* 17, no. 3 (September 2022): 8.

Latham, Michael. *Modernization as Ideology: American Social Science and "Nation Building" in the Kennedy Era*. Chapel Hill: University of North Carolina Press, 2000.

Lee, Wayne E. *Barbarians and Brothers: Anglo-American Warfare, 1500–1865*. New York: Oxford University Press, 2011.

———. "Mind and Matter—Cultural Analysis in American Military History: A Look at the State of the Field." *Journal of American History* 93, no. 4 (2007): 1116–42. https://doi.org/10.2307/25094598.

———, ed. *Warfare and Culture in World History*. New York: NYU Press, 2020.

Lemnitzer, Lyman L. "Why We Need a Modern Army." *Army* 10, no. 2 (September 1959): 16–21.

Leviero, Anthony. "Air Force Calls Army Nike Unfit to Guard Nation." May 21, 1956.

———. "Air Force Doubts Carriers' Value." *New York Times*, May 20, 1956.

———. "Army Fails to Bar Bomb Testimony." *New York Times*, June 29, 1956.

———. "Military Forces Split by Conflict on Arms Policies." *New York Times*, May 19, 1956.

———. "Radford Seeking 800,000 Man Cut; 3 Services Resist." *New York Times*, July 13, 1956.

Levin, Sheldon G. *Mathematical Models for Prediction of Neuropsychiatric and Other Non-battle Casualties in High Intensity Combat*. Aberdeen Proving Ground, MD: US Army Ballistic Research Laboratory, 1986.

Lewis, Adrian R. *The American Culture of War: The History of US Military Force from World War II to Operation Iraqi Freedom*. New York: Routledge, 2007.

Life Magazine. "Airpower Becomes Supply Power." May 15, 1950.

———. "U.S. Trains More Parachute Troops." May 12, 1941.

———. "U.S. Trains Parachutists." August 19, 1940.

Lincoln, G. A., and Amos A. Jordan Jr. "Limited War and the Scholars." *Military Review* 37, no. 10 (January 1958): 50–60.

Lindsey, Douglas. "No Time for Despair." *Armor* 65 (May–June 1956): 36–43.

Linn, Brian McAllister. *The Echo of Battle: The Army's Way of War*. Cambridge, MA: Harvard University Press, 2007.

———. "Eisenhower, the Army, and the American Way of War." Lecture, Eisenhower Lecture Series at Kansas State University, 2003.

———. *Elvis's Army: Cold War GIs and the Atomic Battlefield*. Cambridge, MA: Harvard University Press, 2016.

———. *Real Soldiering: The US Army in the Aftermath of War, 1815–1980*. Lawrence: University Press of Kansas, 2023.

LoFaro, Guy. *The Sword of St. Michael: The 82nd Airborne Division in World War II.* Cambridge, MA: Da Capo, 2011.

Lovelace, Douglas C., and Thomas-Durrell Young. "Defining US Atlantic Command's Role in the Power Projection Strategy." US Army Strategic Studies Institute, August 1998.

Lowden, John L. *Silent Wings at War: Glider Combat in World War II.* Washington, DC: Smithsonian Institution, 1992.

Lynn, John A. *Battle: A Cultural History of Combat and Culture.* Boulder, CO: Westview, 2003.

———. *Bayonets of the Republic: Motivation and Tactics in the Army of Revolutionary France, 1791–94.* Boulder, CO: Westview, 1996.

MacCoun, Robert J., Elizabeth Kier, and Aaron Belkin. "Does Social Cohesion Determine Motivation in Combat? An Old Question with an Old Answer." *Armed Forces and Society* 32, no. 1 (2005): 1-9. https://www.jstor.org/stable/48608737.

Maddry, Joseph K., et al. "Impact of Prehospital Medical Evacuation (MEDEVAC) Transport Time on Combat Mortality in Patients with Noncompressible Torso Injury and Traumatic Amputations: A Retrospective Study." *Military Medical Research* 5, no. 1 (June 2018). https://doi.org/10.1186/s40779-018-0169-2.

Mansoor, Peter R. *The GI Offensive in Europe: The Triumph of American Infantry Divisions, 1941–1945.* Lawrence: University Press of Kansas, 1999.

———. *Surge: My Journey with General David Petraeus and the Remaking of the Iraq War.* New Haven, CT: Yale University Press, 2013.

Mansoor, Peter R., and Williamson Murray, eds. *The Culture of Military Organizations.* New York: Cambridge University Press, 2019.

Martin, Harold H. "Paratrooper in the Pentagon." *Saturday Evening Post,* August 28, 1954.

Martin, Norman E. "Dien Bien Phu and the Future of Airborne Operations." *Military Review* 36, no. 3 (June 1956): 19–26.

Mataxis, Theodore C., and Seymour L. Goldberg. *Nuclear Tactics.* Harrisburg, PA: Military Service, 1958.

McFarland, Linda. *Cold War Strategist: Stuart Symington and the Search for National Security.* Westport, CT: Praeger, 2001.

McGrath, John J. *The Brigade: A History of Its Organization and Employment in the US Army.* Fort Leavenworth, KS: CSI Press, 2004.

McKenney, Janice E. *Organizational History of Field Artillery 1775–2003.* Washington, DC: CMH, 2007.

McKenney, Stewart L. "SKYCAV Operations during Exercise Sagebrush." *Military Review* 36, no. 3 (June 1956): 12–18.

McKenzie, John D. *On Time, on Target: The World War II Memoir of a Paratrooper in the 82d Airborne.* Novato, CA: Presidio, 2000.

McMahon, Robert. "Airmobile Operations." *Military Review* 33, no. 3 (June 1959): 28–35.

McNamara, Robert S. *The Essence of Security: Reflections in Office.* New York: Harper & Row, 1968.

Megellas, James. *All the Way to Berlin: A Paratrooper at War in Europe*. New York: Presidio, 2003.

Michaels, Jeffrey H. "Managing Global Counterinsurgency: The Special Group (CI) 1962–1966." *Journal of Strategic Studies* 35, no. 1 (February 2012): 33–61. https://doi.org/10.1080/01402390.2011.592002.

Midgley, John J., Jr. *Deadly Illusions: Army Policy for the Nuclear Battlefield*. Boulder, CO: Westview, 1986.

Miles, Donna. "From Haiti to Afghanistan, 82nd Shows Flexibility." American Forces Press Service, April 30, 2010. https://www.globalsecurity.org/military/library/news/2010/04/mil-100430-afps02.htm.

Millett, Allan R., and Williamson Murray, eds. *Military Effectiveness*. Vol. 1, *The First World War*. Cambridge: Cambridge University Press, 2010.

———. *Military Effectiveness*. Vol. 3, *The Second World War*. New York: Cambridge University Press, 2010.

Mills, Walter. "General Greunther's Headaches." *Collier's*, July 11, 1953.

Mitchell, Ralph. *The 101st Airborne Division's Defense of Bastogne*. Fort Leavenworth, KS: CSI Press, 1986.

Mooney, Richard E. "Army's Explorer Cheers Congress." *New York Times*, February 2, 1958.

Moore, Harold G., and Joseph L. Galloway. *We Were Soldiers Once . . . and Young: Ia Drang—the Battle That Changed the War in Vietnam*. New York: Ballantine Books, 2004.

Moorman, Frank W. "Logistical Problems in Future Warfare." *Military Review* 30, no. 4 (July 1950): 3–10.

Mrazek, James E. *Airborne Combat: The Glider War / Fighting Gliders of World War II*. Mechanicsburg, PA: Stackpole Books, 2011.

Murphy, Jack. "82nd Paratroopers Forward Deployed near Washington DC." *Connecting Vets*, June 2, 2020. https://www.audacy.com/connectingvets/articles/82nd-paratroopers-forward-deployed-near-washington-dc.

———. "New Details about the 82nd Military Deployment to D.C." *Connecting Vets*, June 11, 2020. https://www.audacy.com/connectingvets/articles/new-details-about-the-82nd-military-deployment-to-dc.

Murphy, Robert M. *No Better Place to Die: The Battle for La Fière Bridge*. Havertown, PA: Casemate, 2009.

Murray, Williamson. *Adaptation in War: With Fear of Change*. New York: Cambridge University Press, 2011.

Muth, Jörg. *Command Culture: Officer Education in the U.S. Army and the German Armed Forces, 1901–1940, and the Consequences for World War II*. Denton: University of North Texas Press, 2011.

Myers, Meghann. "Earning It: A Complete History of Army Berets and Who's Allowed to Wear Them." *Army Times*, November 19, 2017. https://www.armytimes.com/news/your-army/2017/11/19/earning-it-a-complete-history-of-army-berets-and-whos-allowed-to-wear-them/.

NATO Defense College, NDC Conference Report. "The Future of Airborne Forces in NATO." Research Division, NATO Defense College, Rome, Italy, July 2013.

New York Times. "Army Building Up Own Airlift Force." November 16, 1952.

——. "Army's New Guided Missile Takes Off." September 20, 1951.

——. "Gavin Retires, Backs Atomic Tests." April 1, 1958.

——. "Gavin Urged as Senator." January 8, 1958.

——. "Missile Unit Set Up." January 18, 1958.

——. "New Atom Missile Ready for Troops." April 18, 1954.

——. "Paratroopers in Algeria Irked by Political Weakness in Paris." May 16, 1958.

——. "President Pays Honor to Retiring General Ridgway." June 29, 1955.

——. "Tank Leader Tells of Bastogne Fight." June 25, 1945.

——. "Troops Carry on after Atomic Blast 2.7 Miles Away." September 3, 1957.

——. "U.S. Tanks Smash into Bastogne, Break Siege." December 28, 1944.

——. "Westmoreland Surveys Action." November 20, 1965.

——. "Wounded Pleaded for Bastogne Role." January 3, 1945.

Nichols, K. D. "Atomic Guns." *US News & World Report,* July 10, 1953.

Nix, Jack P. "505th Parachute Infantry Regiment (a Legacy of Lessons)." US Army War College Military Studies Program paper, Carlisle Barracks, PA, 1989.

Nordyke, Phil. *All American, All the Way: The Combat History of the 82nd Airborne Division in World War II.* St. Paul, MN: Zenith, 2005.

Norris, John. "Gen. Gavin Would End Joint Staff." *Washington Post and Times Herald,* December 14, 1957.

Nye, Logan. "How the 'Little Groups of Paratroopers' Became Airborne Legends." We Are the Mighty, April 2, 2018. https://www.wearethe mighty.com/articles/how-the-little-groups-of-paratroopers-became-airborne-legends/.

O'Connell, Aaron B. *Underdogs: The Making of the Modern Marine Corps.* Cambridge, MA: Harvard University Press, 2012.

Olson, Wyatt. "US Army Alaska to Be Reflagged as Airborne Division amid Surge in Troop Suicides." *Stars and Stripes,* May 6, 2022. https://www.stripes.com/theaters/us/2022-05-05/us-army-alaska-11th-airborne-division-arctic-strategy-5910999.html.

Osgood, Kenneth A. *Total Cold War: Eisenhower's Secret Propaganda Battle at Home and Abroad.* Lawrence: University Press of Kansas, 2006.

Osgood, Robert. *Limited War: The Challenge to American Strategy.* Chicago: University of Chicago Press, 1957.

Ostlund, William B. "The Largest Air Assault in History." *The Spear* (podcast), produced by John Amble, January 29, 2020. 29:45. https://the-spear.castos.com/episodes/the-largest-air-assault-in-history.

Owen, Robert C. *Air Mobility: A Brief History of the American Experience.* Washington, DC: Potomac Books, 2013.

Palmer, Bruce, Jr. *Intervention in the Caribbean: The Dominican Crisis of 1965.* Lexington: University Press of Kentucky, 1989.

Palmer, Robert J., Bell I. Wiley, and William R. Keast. *The Procurement and Training of Ground Combat Troops.* Washington, DC: GPO, 1948.

Patton, George S. *War as I Knew It.* Boston: Houghton Mifflin, 1947.

Phillips, R. Cody. *Operation Just Cause: The Incursion into Panama.* Center for Military History Publication 70–85–1. Washington, DC: GPO, 2004.

Piasecki, Eugene G. "The Knollwood Maneuver: The Ultimate Test." *Veritas* 4, no. 1 (2008): 54–63. https://arsof-history.org/articles/pdf/v4n1_knollwood.pdf.

Pickrell, Ryan. "The US Military Moved 1,600 Soldiers into Positions outside the Nation's Capital and Has Them on Alert to Respond to Protests If Necessary." Business Insider, June 2, 2020. https://www.businessinsider.com/pentagon-1600-troops-on-alert-outside-dc-for-protest-response-2020-6?op=1.

Pierson, Albert. "Airborne Operations." *Army Information Digest* 9, no. 7 (July 1954): 20–30.

Prados, John. *Vietnam: The History of an Unwinnable War, 1945–1975.* Lawrence: University Press of Kansas, 2009.

Preble, Christopher A. *John F. Kennedy and the Missile Gap.* DeKalb: Northern Illinois University Press, 2004.

Price, Jay. "82nd Airborne Division Celebrates 100 Years." NPR, August 24, 2017. https://www.npr.org/2017/08/24/545757890/82nd-airborne-division-celebrates-its-100th-anniversary.

Public Papers of John F. Kennedy: 1961. Washington, DC: GPO, 1961.

Raff, Edson D. *We Jumped to Fight.* New York: Eagle Books, 1944.

Rathbone, A. D., IV. *He's in the Paratroops Now.* New York: McBride, 1943.

Raymond, Jack. "Thor Is Selected over the Jupiter." *New York Times*, September 29, 1958.

Redmon, Coates. *Come as You Are: The Peace Corps Story.* New York: Harcourt Brace Jovanovich, 1986.

Reinhardt, George C., and William R. Kintner. *Atomic Weapons in Land Combat.* Harrisburg, PA: Military Service, 1953.

Reston, James. "President Asks More Missiles, Further Aid, Pentagon Unit." *New York Times*, January 10, 1958.

Ricks, Thomas. *The Generals: American Military Command from World War II to Today.* New York: Penguin Books, 2012.

Ridgway, Matthew B. "The Army's Role in National Defense." *Army Information Digest* 9, no. 5 (May 1954): 21–25.

——. "How Europe's Defenses Look to Me." *Saturday Evening Post*, October 10, 1953.

——. *The Korean War.* New York: Doubleday, 1967.

Ridgway, Matthew B., as told to Harold H. Martin. *Soldier: The Memoirs of Matthew B. Ridgway.* New York: Harper, 1956.

Riley, Rachel. "Commander of 82nd Airborne Division to Deploy to Afghanistan." *Fayetteville Observer*, August 18, 2021. https://www.fayobserver.com/story/news/2021/08/17/pentagon-82nd-airborne-division-role-afghanistan-maj-gen-christopher-donahue-military-biden-taliban/8165782002/.

——. "Fort Bragg Soldiers Deploy to Support NATO in Europe amid Russian Threat to Ukraine." *Fayetteville Observer*, February 3, 2022. https://www.

fayobserver.com/story/news/2022/02/03/fort-bragg-soldiers-deploy-europe-russia-presence-ukraine-border-putin-us-military/6646815001/.

Rockis, Joseph. "Reorganization of Army Ground Forces during the Demobilization Period." AGF Demobilization Study #3. Fort Monroe, VA, 1948.

Rogers, Clifford J., ed. *The Military Revolution Debate: Readings on the Military Revolution of Early Modern Europe*. New York: Routledge, 1995.

Romjue, John L. *American Army Doctrine for the Post–Cold War*. Washington, DC: CMH, 1997.

Rosen, Stephen Peter. *Winning the Next War: Innovation and the Modern Military*. Ithaca, NY: Cornell University Press, 1994.

Rottman, Gordon. *U.S. Army Airborne, 1940–90*. London: Osprey, 1990.

Royal, Everett C. "The Team of Mobile Warfare: Armor and Airborne." *Armor* 65, no. 2 (March–April 1955): 4–6.

Ryan, John, and Tony Dokoupil. "In Afghanistan's 'Valley of Death' a Medevac Team's Miracle Rescue." *Newsweek*, November 5, 2012. https://www.newsweek.com/afghanistans-valley-death-medevac-teams-miracle-rescue-63779.

Sampson, Francis L. *Look Out Below! A Story of the Airborne by a Paratrooper Padre*. Sweetwater, TN: 101st Airborne Division Association, 1989.

Samuels, Gertrude. "Ridgway—Three Views of a Soldier." *New York Times Magazine*, April 22, 1951.

Sayre, Edwin M. "The Operations of Company A, 505th Parachute Infantry (82d Airborne Division), Airborne Landings in Sicily, 9–24 July 1943 (Sicily Campaign) Personal Experiences of a Company Commander." Fort Benning, GA, 1947.

Scales, Robert H. *Certain Victory: The U.S. Army in the Gulf War*. Washington, DC: Potomac Books, 1997.

Schadlow, Nadia. *War and the Art of Governance: Consolidating Combat Success into Political Victory*. Washington, DC: Georgetown University Press, 2017.

Schein, Edgar H., and Peter Schein. *Organizational Culture and Leadership*. Hoboken, NJ: John Wiley, 2016.

Scheips, Paul J. *The Role of Federal Military Forces in Domestic Disorders 1945–1992*. Washington, DC: CMH, 2005.

Schenectady Gazette. "Approve New Emblem for Parachutists." July 2, 1941.

School, Jeff. "Why a 2-Star General Was the Last American Service Member to Leave Afghanistan." Task & Purpose, August 31, 2021. https://taskandpurpose.com/news/army-general-last-soldier-leave-afghanistan/.

Schrader, Charles R. *The First Helicopter War: Logistics and Mobility in Algeria, 1954–1962*. Westport, CT: Praeger, 1999.

Schrepel, Walter A. "Paras and Centurions: Lessons Learned from the Battle of Algiers." *Peace and Conflict: Journal of Peace Psychology* 11, no. 1 (July 2005): 71–89. https://doi.org/10.1207/s15327949pac1101_9.

Schuster, Alvin. "Gen. Gavin, Missile Aide, to Quit; Criticized Joint Chiefs System." *New York Times*, January 5, 1958.

Schwarzkopf, H. Norman, with Peter Petre. *It Doesn't Take a Hero*. New York: Bantam Books, 1992.

Scott, Ridley, dir. *Black Hawk Down*. Sony Pictures, 2001.

Seelinger, Matthew. "The M28/M29 Davy Crockett Nuclear Weapon System." Army Historical Foundation. https://armyhistory.org/the-m28m29-davy-crockett-nuclear-weapon-system/.

Sepp, Kalev I. "The Pentomic Puzzle: The Influence of Personalities and Nuclear Weapons on U.S. Army Organization 1952–1958." *Army History* 51 (Winter 2001): 1–13.

Shallet, Sidney. "Arnold Reveals Secret Weapons, Bomber Surpassing All Others." *New York Times*, August 18, 1945.

Sherburne, T. L. "Reorganizing the 101st Airborne Division: An Interim Report." *Army Information Digest* 12, no. 6 (June 1957): 12–23.

Sherry, Michael S. *The Rise of American Air Power: The Creation of Armageddon*. New Haven, CT: Yale University Press, 1987.

Sicard, Sarah. "The Mysterious Origins of 'HOOAH,' the Army's Beloved Battle Cry." Task & Purpose, October 5, 2017. https://taskandpurpose.com/history/mysterious-origins-hooah-armys-beloved-battle-cry/.

Simpkins, J. D. "Army to Honor 1st Cavalry Division with New Unis against Rival Navy." *Military Times*, December 6, 2019. https://www.militarytimes.com/off-duty/military-culture/2019/12/06/army-uniforms-honor-1st-cavalry-division-for-navy-rivalry-game/.

Soffer, Jonathan. *Matthew B. Ridgway: From Progressivism to Reaganism, 1895–1993*. Westport, CT: Praeger, 1998.

Sorley, Lewis. *Honorable Warrior: General Harold K. Johnson and the Ethics of Command*. Lawrence: University Press of Kansas, 1998.

———. *Thunderbolt: General Creighton Abrams and the Army of His Times*. New York: Simon & Schuster, 1992.

Speranza, Vincent J. *NUTS! A 101st Airborne Division Machine Gunner at Bastogne*. Atlanta: Deeds, 2014.

Spielberg, Steven, and Tom Hanks, producers. *Band of Brothers*. HBO Entertainment, 2001.

Spiller, Roger J. *"Not War but Like War": The American Intervention in Lebanon*. Fort Leavenworth, KS: CSI Press, 1981.

Spokane Daily Chronicle. "Weapons' Values to Be Appraised." December 15, 1948.

Stanton, Shelby L. *Anatomy of a Division: 1st Cav in Vietnam*. Novato, CA: Presidio, 1987.

———. *Vietnam Order of Battle*. Mechanicsburg, PA: Stackpole Books, 2003.

Starr, Chester G. *From Salerno to the Alps: A History of the Fifth Army, 1943–1945*. Washington, DC: Infantry Journal Press, 1948.

Stars and Stripes. West Point's Army-Navy Game Uniforms to Honor 82nd Airborne." December 6, 2016. https://www.stripes.com/sports/west-point-s-army-navy-game-uniforms-to-honor-82nd-airborne-1.442946.

Steigerwald, David. *The Sixties and the End of Modern America*. New York: St. Martin's, 1995.

Stevens, Austin. "Planes Fall Short of Airlift Needs." *New York Times*, May 2, 1950.

Stewart, Richard W. *Operation Urgent Fury: The Invasion of Grenada, October 1983*. Washington, DC: CMH, 2009.

———. *The United States Army in Afghanistan: Operation Enduring Freedom, October 2001–March 2002*. Washington, DC: CMH, 2004.

Stouffer, Samuel A., et al. *The American Soldier*. Vol. 2, *Combat and Its Aftermath*. Studies in Social Psychology in World War II. Princeton, NJ: Princeton University Press, 1949.

Stromseth, Jane E. *The Origins of Flexible Response: NATO's Debate over Strategy in the 1960s*. London: Macmillan, 1988.

Taaffe, Stephen R. *Marshall and His Generals: U.S. Army Commanders in World War II*. Lawrence: University Press of Kansas, 2011.

Taylor, John M. *General Maxwell Taylor: The Sword and the Pen*. New York: Doubleday, 1989.

Taylor, Maxwell D. *Swords and Plowshares*. New York: W. W. Norton, 1972.

———. *The Uncertain Trumpet*. New York: Harper & Bros., 1959.

Thompson, Jack. "Lack of Planes Delays Training of Paratroops." *Chicago Daily Tribune*, July 14, 1941.

Thornton, Ron. "Getting It Wrong: The Crucial Mistakes Made in the Early Stages of the British Army's Deployment to Northern Ireland (August 1969 to March 1972)." *Journal of Strategic Studies* 30, no. 1 (February 1, 2007): 73–107. https://doi.org/10.1080/01402390701210848.

Tolson, John J. *Airmobility 1961–1971*. CMH Publication 90-4. Washington, DC: CMH, 1973.

Trauschweizer, Ingo. "Berlin Commander: Maxwell Taylor at the Cold War's Frontlines, 1949–51." *Cold War History* 21, no. 1: 52–53.

———. *The Cold War U.S. Army: Building Deterrence for Limited War*. Lawrence: University Press of Kansas, 2008.

———. *Maxwell Taylor's Cold War: From Berlin to Vietnam*. Lexington: University Press of Kentucky, 2019.

Turner, Stansfield. "Towards a New Defense Strategy." *New York Times*, May 10, 1981.

US House of Representatives. *Army Appropriations Subcommittee, Hearings for FY 1955*, 83rd Congress, 2nd Session, 1954.

———. *Committee on Armed Services, Hearings before Committee on Armed Services*, 85th Congress, 1st Session, 1956.

———. *House Committee on Appropriations, Department of Defense Appropriations for 1983*, 97th Congress, 2nd Session.

US Senate. *Defense Appropriations Subcommittee, Defense Appropriations Hearings for FY 1956*, 84th Congress, April 4–June 6, 1955.

———. *Inquiry into Satellite and Missile Programs*, Committee on Armed Services, 85th Congress, 1st Session, December 13, 1957.

———. *Nomination Hearings before the Committee on Armed Services*, 83d Congress, 1st Session, 1953.

———. *Study of Airpower*. Hearings before the Subcommittee on the Armed Services, 84th Congress, 2nd Session, April 16–June 1, 1956.

US War Department. FM 31-30, *Tactics and Technique of Air-borne Troops*. Washington, DC: GPO, 1942.

———. *Reports of the General Board U.S. Forces, European Theater of Operations*. 1945.

———. Training Circular 113. October 9, 1943.

Valliere, John E. "Disaster at Desert One: Catalyst for Change." *Parameters* 22, no. 1 (Autumn 1992): 69–82.

Vanderpool, Jay D. "We Armed the Helicopter." *Army Aviation Digest* 17, no. 6 (June 1971): 2–6, 24–29.

Van Gennep, Arnold. *The Rites of Passage*. Translated by M. B. Visedom and G. L. Caffe. Chicago: University of Chicago Press, 1960.

Venable, Heather P. *How the Few Became the Proud: Crafting the Marine Corps Mystique, 1874–1918*. Annapolis, MD: Naval Institute Press, 2019.

Wade, Gary H. *Rapid Deployment Logistics: Lebanon, 1958*. Fort Leavenworth, KS: CSI Press, 1984.

Walkowicz, T. F. *Future Airborne Armies: A Report Prepared for the AAF Scientific Advisory Group*. Wright Field, OH: Air Material Command, September 1945.

Wallace, Randall, dir. *We Were Soldiers*. Los Angeles: Paramount Pictures, 2002.

"War and Peace in the Nuclear Age: Bigger Bang for the Buck; an Interview with James Gavin, 1986." February 25, 1986. GBH Archives, Boston.

Warren, Jason W., ed. *Drawdown: The American Way of Postwar*. New York: NYU Press, 2016.

Washington Post. "A Flotilla of Army Helicopters Joins Attack on Karbala." March 29, 2003. https://www.washingtonpost.com/archive/politics/2003/03/29/a-flotilla-of-army-helicopters-joins-attack-on-karbala/3b318b62-7251-406f-b762-2cf0e8822df1/.

Weapon Systems Evaluation Group. "A Historical Study of Some World War II Airborne Operations." WSEG Staff Study No. 3, 1951.

Wellman, William A., dir. *Battleground!* Metro-Goldwyn-Mayer, 1949.

Wermuth, Anthony L. "Modernization-Minus." *Army* 9, no. 3 (October 1958): 26–34.

Westmoreland, William C. *A Soldier Reports*. New York: Doubleday, 1976.

Weston, Leonard C. *Project Management of the Davy Crockett Weapons System, 1958–1962*. Rock Island, IL: Rock Island Historical Branch, 1964.

Wheeler, Earle G. "Army Moves toward Mobility." *Army Information Digest*, February 1964, 34–35.

———. "Strategic Mobility." *Army Information Digest* 12, no. 1 (January 1957): 2–12.

White, Theodore H. "An Interview with General Gavin . . . Tomorrow's Battlefield." *Army Combat Forces Journal* 5 (March 1955): 20–23.

Williams, R. F. M. "Bring Back the Sightseeing Sixth: The Case for an Arctic Division." Modern War Institute, December 14, 2021. https://mwi.usma.edu/bring-back-the-sightseeing-sixth-the-case-for-an-arctic-division/.

———. "Integrating Army Capabilities into Deterrence: The Early Cold War." *Parameters* 53, no. 5 (Winter 2023): 69–82. https://doi.org/10.55540/0031-1723.3260.

———. "'Our Problem Children': Masculinity and Its Discontents in American Parachute Units in World War II." *Journal of Military History* 87, no. 3 (July 2023): 675–702.

Wilson, John B. *Maneuver and Firepower: The Evolution of Divisions and Separate Brigades*. Washington, DC: CMH, 1998.

Wilson, Peter H. "Defining Military Culture." *Journal of Military History* 72 (2008): 18.

Winkie, Davis. "82nd Airborne's Special Relationship with D-Day Village Endures Virtually amid Pandemic." *Army Times*, December 16, 2020. https://www.armytimes.com/news/your-army/2020/12/15/82nd-airbornes-special-relationship-with-d-day-village-endures-virtually-amid-pandemic/.

——. "Why the 82nd Airborne Is Directing Airfield Security for Afghanistan Evacuation." *Army Times*, August 17, 2021. https://www.militarytimes.com/flashpoints/afghanistan/2021/08/17/why-the-82nd-airborne-is-directing-airfield-security-for-afghanistan-evacuation/.

Winslow, Donna. *The Canadian Airborne Regiment in Somalia: A Socio-cultural Inquiry*. Ottawa: Canadian Government, 1997.

Wintersteen, Joseph O., Jr. "Helicopterborne Operations." *Infantry Journal* 47, no. 2 (April 1957): 20–33.

Wright, Robert K. "Airborne Forces and the American Way of War." *Army History* 72 (Summer 2009): 36–48. https://www.jstor.org/stable/26298712.

Wurst, Spencer F., and Gayle Wurst. *Descending from the Clouds: A Memoir of Combat in the 505 Parachute Infantry Regiment, 82nd Airborne Division*. Havertown, PA: Casemate, 2004.

Wyman, Willard G. "Let's Get Going on Our New Combinations for Combat." *Army* 6 (July 1956): 38–43.

Yarborough, William P. *Bail Out over North Africa: America's First Combat Parachute Missions, 1942*. Williamstown, NJ: Phillips, 1979.

Yockelson, Mitchell. *The Paratrooper Generals: Matthew Ridgway, Maxwell Taylor, and the American Airborne from D-Day through Normandy*. Guilford, CT: Stackpole Books, 2020.

Zaloga, Steven J. *Inside the Blue Berets: A Combat History of Soviet and Russian Airborne Forces, 1930–1995*. Novato, CA: Presidio, 1995.

INDEX